The Future Factor

The Future Factor

Factor

The Five Forces Transforming Our Lives and Shaping Human Destiny

Michael G. Zey, Ph.D.

McGraw-Hall

New York   San Francisco   Washington, D.C.   Auckland   Bogotá
Caracas   Lisbon   London   Madrid   Mexico City   Milan
Montreal   New Delhi   San Juan   Singapore
Sydney   Tokyo   Toronto

Library of Congress Cataloging-in-Publication Data

Zey, Michael G.
   The future factor : the five forces transforming our lives and shaping our destiny /
Michael G. Zey.
      p. cm.
Includes bibliographical references and index.
ISBN 0-07-134305-9 (alk. paper)
1. Social prediction.   2. Technological forecasting.   I. Title.

HN17.5 .Z48   2000
303.49—dc21   00-055412

McGraw-Hill

A Division of The McGraw·Hill Companies

1 2 3 4 5 6 7 8 9 0   AGM/AGM   0 9 8 7 6 5 4 3 2 1 0

ISBN 0-07-134305-9

This book was set in Garamond by Binghamton Valley Composition.

Printed and bound by Quebecor World/Martinsburg.

McGraw-Hill books are available at special quantity discounts to use as premiums and
sales promotions, or for use in corporate training programs. For more information,
please write to the Director of Special Sales, Professional Publishing, McGraw-Hill,
Two Penn Plaza, New York, NY 10121-2298. Or contact your local bookstore.

This book is printed on acid-free paper.

To George, my father, and Aunt Kay, who brought me wisdom, love, and joy

Contents

Preface

The fact that you are perusing a book entitled *The Future Factor* reveals that you are a person who cares deeply about the future of the species and the planet. You want to know about the trends that are transforming your world, and how you can help make that world a better place for you, your children, and the generations to come. You believe that humanity can solve the problems that confront it. You are, at heart, an optimist.

The Future Factor, an exuberantly upbeat book about the near-term and long-range prospects of the human species, the planet, and the universe, is written for you.

This book can be read on two different though ultimately related levels. On the first level, this book describes how breakthroughs in science and technology will impact our economy, society, and future generations. Throughout I describe the breathtaking innovations in fields such as biotechnology, computing, robotics, medicine, energy development, and space technology that will change forever the way we live. You will learn how we will double the average human life span, colonize Mars and beyond, and in our lifetimes witness the emergence of a world of unlimited abundance and prosperity.

This book goes on to tell a second, equally fascinating story, about the long-term future of the human species. I describe cutting-edge theories and research in cosmology, physics, and other fields that offer a startling, yet inspiring, vision of human destiny that shatters standard concepts of the relationship of humankind to the cosmos.

I wrote this book for people who want a grounding in the future and have a desire to understand emerging social and technological trends in a clear and objective way. These are people who want to become literate in the science and technology of their times and understand how breakthroughs in genetics, computers, and robotics will affect their lives, their careers, and the world around them.

In addition, this book has practical applications—it describes how scientific and technological advancements will impact the economy, society, and business in general. I believe the information here will be of great value to a wide range of businesspeople: CEOs, middle managers, marketing professionals, corporate strategic and tactical planners, and those in the early stages of their careers. Even investors will discover a wealth of useful insights about the emerging economic trends and new companies to be watched. If you fall into any of these categories you can use this knowledge to help you make a wide variety of business and career decisions.

In this book you will learn about significant innovations and inventions unfamiliar to the vast majority of the public. Much of this information comes my way via my contacts in the business, academic, military, governmental, and political worlds. I also maintain connections to the "techno-underground" of independent scientists, political activists, and academics working in fields such as space exploration, macroengineering, cryonics, and bioengineering. In addition, my consulting and lecturing activities give me access to intelligence about upcoming breakthroughs in computers, engineering, and bioscience.

I believe that this book contains several important lessons for all readers. First, it demonstrates that humanity has a bright future, a destiny, as it were. Second, it illustrates that humanity is just in its infancy and that our best days are ahead of us. Third, I show how science and technology can be used for the good of humanity, the planet, and ultimately the universe. In addition, I assert that the human species represents a positive development in the evolution of the cosmos.

The Future Factor will provide you with a unique way to think about the future, humanity's role in it, as well as the social and technological influences that are buffeting our society as we enter the third millennium.

Moreover, I can promise that once you have read this book, you will feel heartened by the future that awaits the human species. My sincerest hope is that this volume also encourages you to participate in our growing efforts to forge a pathway to a better world!

Michael G. Zey, Ph.D.
Morristown, New Jersey

Acknowledgments

While writing such a book is a joyful and exhilarating experience, it can also prove to be an arduous task. I could not even hope to complete such a task without the assistance and patient cooperation of a host of individuals and groups.

I first would like to extend the warmest gratitude to Ms. Jennifer Occhipinti, my tireless researcher for the better part of two years. I will always appreciate the enthusiasm she demonstrated for the project, and the energy she applied to the research. More importantly, I genuinely enjoyed the many discussions we had regarding a wide range of issues and topics related to *The Human Future*. Her insights into the material were invaluable.

Special kudos and accolades go to Cindy Zey, who demonstrated superhuman tolerance and patience during the period in which I wrote this book. I also appreciated her insights into the material, help in editing the manuscripts, and advice on the book's structure. And most of all, I thank her for all the needed encouragement throughout this time.

Researchers Maria Johnson and Meredith Griffin also provided valuable help in preparing this manuscript. And thank you Ms. Carolyn Keating, for all the assistance you have provided me over these years.

My agent, Robert Tabian, has supported my efforts through my career. I hope we continue to successfully pursue future together.

I want to especially thank Mary Glenn, my editor at McGraw-Hill who had the faith in this project to ensure that it get published. Thank you also Amy Murphy, for providing much assistance during the writing process, and Jane Palmieri, for her careful overseeing of the book's production process.

Many thanks go to Michael, Carol, and Katie Aloisi, for their friendship and stimulating intellectual conversation that served as a catalyst to the writing process; my family, for their love and support; and Franny and Zooey Zey, for always being there for me.

Also, I would like to recognize Montclair State University, the School of Business, and my colleagues in my department for providing me the support I needed to bring this book to a successful conclusion. I especially appreciate the insights provided by Montclair's Dr. Carl Rodrigues.

Prologue
Creating the Human Future

In *Seizing the Future: The Dawn of the Macroindustrial Era,* I charted the recent monumental and breathtaking advances in computers, biotechnology, aerospace, engineering, and transportation that will help us reshape the planet and fundamentally improve the human condition.

In *The Future Factor,* I continue to explore the myriad ways that the human species is commandeering nature and extending its powers across the planet and beyond. Over the past few years the rate of human achievement in all scientific fields has accelerated dramatically. We are transforming the human body; tinkering with our genetic arrangement to make ourselves smarter, faster, and healthier; and then developing ways to clone the final product. We enhance the very functioning of the brain, and implant that human brainpower into our machines. We unify the human family by developing global communication systems, such as the Internet. As our species extends our control over this planet, we simultaneously prepare ourselves for extraterrestrial habitation by shaping and transforming terrestrial landscapes. We design a new generation of rockets that can transport us to distant spheres at one-third the speed of light. At the same time, we probe the innermost recesses of nature through such exotic fields as nanotechnology.

In *The Future Factor* I will examine the many ways such developments impact the individual, society, and the economy. In addition, I will delve into a much more profound issue, the underlying reasons why our species is feverishly working to advance the planet and ourselves and transform all we encounter.[1] Only when we truly understand the depth and strength of man's overwhelming imperative to grow and progress can we clearly anticipate the future.

At first blush, it would seem that there is little mystery about the impulses driving the human species in this quest: we engage in such productive activities merely to enhance our material condition. We invent technologies that

1

will improve our standard of living and make our lives more pleasant and comfortable. Notice, however, that our species from the earliest periods of prehistory seems compelled not just to survive, but to grow, progress, and enhance itself and its environment. Our species probably could have chosen to not progress beyond relatively humble levels of material development, such as the hunter-gatherer stage we reached thousands of years ago, and still maintained an adequate though inconsequential existence. Yet our species has striven ceaselessly to improve our physical lot. In addition, at each new level, we endeavor to master our environment as well as the physical dynamics governing our universe.

In *The Future Factor*, I describe what has become increasingly apparent to other researchers and myself: Humanity's activities, including the entire scientific and technological enterprise, represent a unified attempt by the species to spread "humanness" to everything we encounter. Over the centuries we have labored to improve planet Earth, and as the third millennium begins, we are preparing to *transform the universe into a dynamic entity filled with life.* We will accomplish this by extending our consciousness, skills, intellect, and our very selves, to other spheres.

I label the sum total of our species' endeavors to improve and change *our* planetary environment and ultimately the universe itself vitalization. *Vitalization*, I contend, is a *force* conditioning human behavior. In fact, I will demonstrate in the course of this book that the drive to *vitalize*, to imbue our planet and eventually the cosmos with consciousness and intelligence, is a primary motivation behind all human productive activity.

How We Will Get There—The Building Blocks of Vitalization

Vitalization is the primary force shaping human behavior. However, in order to pursue vitalization successfully, the human species must master four other forces, what I label the "building blocks of vitalization." These four processes encompass the extraordinary advances in areas such as space, medicine, biogenetics, engineering, cybernetics, and energy.

These four supporting forces are

- Dominionization
- Species Coalescence

- Biogenesis
- Cybergenesis

These forces are in actuality dynamic processes that serve as catalysts in humankind's forward progression to its next levels of development. As we enter the twenty-first century, we are already hard at work building the foundation necessary to complete vitalization. Two of the processes described in Chart 1—*dominionization* and *species coalescence*—could reach a mature stage as early as the end of the twenty-first century. These processes are occurring during the Macroindustrial Era, a concept I introduced in my last book. This era, which began in earnest in the early 1980s, represents a defining moment in human development, the early stages of a qualitative change in the relationship between humanity and cosmos. *Biogenesis* and *cybergenesis* are in earlier stages of development.

Chart 1
The Building Blocks of Vitalization

Dominionization	The achievement of control over several key physical dimensions. This includes the enhanced ability to control matter, develop innovative energy sources, reshape the topography and geography of planets and other spheres, and control many of the basic phenomena of the physical universe, including electromagnetism and the behavior of the atom.
Species Coalescence	The process whereby humanity achieves total unity—physically, culturally, and functionally. It includes the development of a global transportation grid, a global communications network, and a universal production system.
Biogenesis	The improvement of the human physical shell through the application of advanced genetic technologies, biotechnology, and a variety of breakthroughs in fields such as cloning and nanotechnology.
Cybergenesis	The interconnection of man and machine to advance human evolution, specifically through the enhancement and expansion of the functioning of the brain through utilization of cybernetics, biofeedback, and other computer-based technologies. Also included is the use of the computer as a surrogate memory, systems simulator, and high-speed mathematical calculator.

I will identify here the critical role that each force plays in the achievement of vitalization, and elaborate on this issue throughout the book. Certainly,

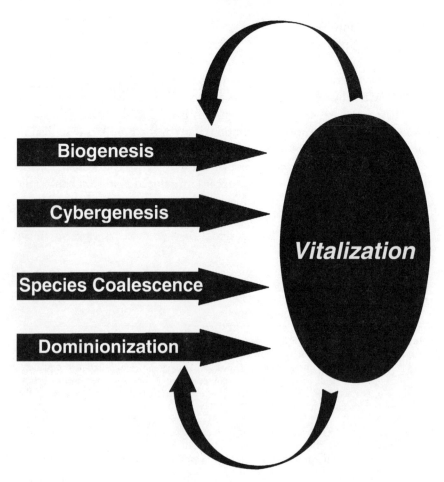

Figure 1

the details of humanity's current efforts to achieve vitalization as it enters the twenty-first century are nearly as exciting as the vision of our destiny itself. (The interplay among these five forces is illustrated in Figure 1.)

Dominionization

The term *dominionization* refers to the process whereby humankind establishes control over several key aspects of its physical universe. With each passing decade we enhance our ability to manipulate matter, reshape the planet, develop innovative energy sources, and control fundamental aspects of the physical universe, such as the atom and electromagnetism. Someday, we will learn to influence weather patterns and climate.

In a host of ways dominionization helps humanity vitalize the planet and eventually the universe. As we master the basic dynamics of nature, we are more able to shepherd the evolution of our planet as well as others. As we develop novel and powerful forms of energy we can rocket from one sphere to another. Moreover, by improving our already formidable skills in moving mountains and creating lakes we will be better able to change both the topography and geography of other planets.

Examples of dominionization abound. Major macroengineering projects attest to man's ability to transform the very surface of Earth. By constructing man-made lakes we will be able to live in previously uninhabitable areas such as interior Australia. Shimizu Corporation envisions a subterranean development called Urban Geo Grid—a series of cities linked by tunnels—accommodating half a million people. In the Macroindustrial Era, we will redefine the concept of "bigness" as we dot Earth's landscape with immense architectural structures. Takenaka, a Japanese construction firm, has proposed "Sky City 1000," a 3000-foot tower, to be built in Tokyo. Another firm, Ohbayashi, plans to erect a 500-story high-rise building featuring apartments, offices, shopping centers, and service facilities.[2]

We will establish dominion over the very heart of physical matter itself. Through *nanotechnology* our species will attain control over the atom and its tiniest components. Such control will enable us to effortlessly "macromanufacture" from the bottom up, one atom at a time, any material object, enabling us to permanently eradicate age-old problems such as scarcity and poverty. We will also establish *dominion* over our physical realm by mastering the energy production process. We are on the verge of developing a cheap, accessible form of nuclear fusion for general use, and various companies and government agencies are seriously experimenting with exotic phenomena such as electromagnetism to explore its possible application to energy production. We will travel to the moon, planets, and asteroids to mine exotic new forms of energy. Organizations such as the Space Studies Institute in Princeton, New Jersey, are drawing up the blueprints for a Solar Power Satellite that will sit in geostationary orbit above the equator, collect cheap and abundant solar energy, and beam it down to Earth in microwave form for land-based energy production and consumption.[3]

Species Coalescence

In the Macroindustrial Era, humanity will also pursue another process critical to the achievement of vitalization. Species coalescence refers to the sequence through which humanity achieves total unity—physically, culturally, and functionally.

Species coalescence empowers us to achieve vitalization in a number of ways. On the purely functional level, such coalescence makes it more likely that all members and groups across the globe will work together, pooling their diverse skills and talents, to facilitate the vitalization process. In addition, coalescence enables us to develop a sense of membership in a common human family, not unlike the sense of common identity shared by members of a clan, village, or a neighborhood community. People who partake of a common identity are more likely to commit to collective goals, in this case vitalization.

Specifically, mankind is accelerating species coalescence through the development of a global transportation grid, the universal communications network, and other mechanisms. The global transportation grid includes the hyperplane, which promises to reduce the New York-to-Tokyo trip to 2 hours; smart roads to speed automobiles; and supertunnels such as the Chunnel. Construction of a global high-speed rail system will prove crucial to the achievement of species coalescence. In 1998, the United States finally passed legislation that would release millions of dollars for the development of a magnetic levitation train line. Construction of such a 310-mile per hour train could begin as early as 2001.[4]

Governments and corporations are partnering in the construction of a global power grid that would link Europe, North America, and Asia. With this grid in operation Siberia could send North America power over a link across the Bering Strait. In addition, a number of tunnel, bridge, and causeway projects will accelerate the species coalescence process. A Gibraltar crossing may soon connect Europe and Africa. Moreover, various countries are still planning to build within the decade the proposed Bering Strait crossing, which would link Asia with North America.[5]

The universal communication network of images, voice, and data is made possible by satellites, fiber optics, and other advanced technologies. The Internet, still in its infancy, is only a harbinger of things to come. Hastening this societal convergence is the fact that previously unconnected nations and communities are becoming integrated into the worldwide production-consumption web as manufacturers, designers, inventors, and consumers.

Space colonization will compel countries to form partnerships to build space stations and eventually explore and colonize Mars. For example, in 1998, Japan launched a powerful vehicle, the M-5 four-stage rocket, that would reach Mars by 2000 at the latest. Japan was particularly interested in finding traces of water beneath the Martian surface and other signs of early life. The United States, Canada, Sweden, and Germany contributed 4 of the 14 experiments on board the craft.[6]

Biogenesis

Although remarkable, these triumphs of human ingenuity represent only part of the process by which humanity is vitalizing the cosmos. As we enter the next century, humanity is feverishly working toward assuming control of its very physical development and long-term evolutionary advancement. I label this process *biogenesis*: the modification, enhancement, and in some cases transformation of the human body.

To achieve vitalization we must assume control over our physical selves. Vitalization will require an "enhanced" version of the human being—smarter, more adroit, more creative. Through such techniques as genetic manipulation, cloning, and other forms of physical reconfiguring of the body human we will create that "new and improved" human. Moreover, since we will be traveling to other planets and distant galaxies, we must learn how to modify the body, adapt it to these different environments. Also, protracted processes such as vitalization require human physical beings that are durable; they must live decades or centuries longer than they do now.

We will achieve biogenesis by developing a host of new technologies. Scientists at the Geron Corporation and the University of Colorado at Boulder claim they have found the immortality gene, the gene that controls the aging process in humans.[7] They boast that by tinkering with this gene, they soon might be able to equip the human body with a set of instructions to simply stop aging. The magical new science, nanotechnology, will allow us to shape and redesign the human body as we see fit. It opens the possibility of creating wholly new organs to help us adapt to new environments. We might even develop nanomachines that cruise through a patient's body and fight viruses, including AIDS. Companies such as Advanced Tissue Sciences, Inc., and Organogenesis have developed tissue regeneration and tissue-engineering techniques that will make the human body resistant to diseases such as Alzheimer's and cystic fibrosis, and could help the body regenerate missing parts. Another development, cloning, will allow us to replicate body types resistant to diseases and adaptable to a variety of environments (including outer space).[8] Also, bionics will further species development. Researchers in the United States, at Johns Hopkins University and University North Carolina at Chapel Hill, as well as those in Japan, are working on an artificial retina that will allow blind persons to be able to detect motion and light and eventually simulate sight.

It is no wonder that as we enter the twenty-first century popular periodicals describe a future in which we may expect "bodies by design" and "bioengineered humans" as commonplace occurrences.[9]

Cybergenesis

Throughout the twenty-first century, humanity will engage in another process crucial to its advancement. *Cybergenesis* refers to the incorporation of computers, microchips, and cybernetics into the human evolution process. Robotics, and automation in general, will enhance the functioning of the human brain. The computer will contribute to the human species' growth and development by enhancing the brain's functioning, as a surrogate memory, visualizer, calculator, and decision maker.

To the extent that it makes us smarter, and hence more adaptable, cybernetics becomes an integral part of our achievement of vitalization. Cybergenesis will imbue the species with the brainpower and mental dexterity to perform the computational and conceptual feats required to vitalize planets, including our own.

Soon nanocomputers will be placed inside the brains of humans to enhance memory, thinking ability, visualization, and general sensing. Researchers at British Telecommunications PLC seek to develop a computer that can be implanted in the brain to complement human memory and computational skills. In addition, technologies are at hand that will enable the human brain to "connect" to a computer and download and upload data.

"Brain science" researchers are revealing how technology can help us expand humans' physical and mental abilities and powers. At such sites as the Yale/Veterans Affairs PET Center in West Haven, Connecticut, scientists are diligently developing ways for an amputee's brain to redesign itself and grow new communication paths to reattached limbs. The U.S. government senses the importance of such research into cognitive functioning. As part of the "Decade of the Brain," Congress in the 1990s authorized the "Human Brain," a joint effort of NASA, the National Institute of Mental Health, the National Institute on Aging, the National Science Foundation, and the Office of Naval Research. This project's purpose is to develop technology to alter and improve the operation of the human brain and enlarge people's brain capacity, thereby furthering the process of cybergenesis.

By its very definition, vitalization involves the transference of organic life and human consciousness to other spheres throughout the universe. Therefore, we will examine the role space exploration and colonization play in both the evolution of humanity and the ultimate development and perfection of the universe. Through space exploration and travel we will fundamentally redefine ourselves, from Earth-bound to "extraterrestrial." At that point we will become active participants in the development of the universe as we spread human consciousness and organic life itself throughout a currently dead universe.

Vitalization will also be achieved through *terraformation*, the creation of organic environments on Mars and other barren spheres. This process is being enabled by pioneering work in this nascent field performed by the Engineering Society for Advancing Mobility (SAE) and the Millennial Project, as well as bold new efforts by countries such as China, Japan, and the United States, to colonize planets and vitalize these spheres' terrestrial and climatic environments. (China jolted the world by announcing that it planned to loft two men into orbit on a spacecraft named *Project 921* in 2000.)

I should point out that this does not represent a stage or stepwise process. That is, we do not have to complete biogenesis or dominionization before we begin vitalizing the universe. Rather, we must continue to make progress in each of these dimensions in order to successfully engage in vitalization. Also, each of these processes impacts the other to such an extent that a powerful synergy emerges. For example, advances in the cybergenesis area could accelerate the coalescence of the human species. (See Figure 1.)

I emphasize throughout that our success in meeting our objectives is only partly the result of the plethora of technological and scientific breakthroughs. The confluence of politics and culture and the realities of the local and global economy also play a major role in determining how successful our species will be in achieving its destiny. Humanity's long march toward vitalization has been spearheaded by the pro-progress, pro-species global network of academic, governmental, political, and cultural groups devoted to advancing the species. At the same time, however, social and cultural counterforces are amassing—in academia, grade schools, the media, and the natural and social sciences—that threaten to upend the very progress promised by our scientific developments. I chronicle the "battle for the future" currently being fought in all our institutions over the meaning of the human species, the proper scope of human endeavor, and ultimately the rights of humans on this planet and throughout the universe. The outcome of this battle will largely determine whether we will be able to muster the will to meet our cosmic mandate.

An Emerging Vision of Human Destiny

The first part of this book establishes the fact that humanity indeed is involved in the process of vitalization, the extension of human intelligence and consciousness throughout the planet and to other worlds. The question, of course, is "why are we engaged in this enterprise?"

In Chapter 6, I explore the possibility that our species is guided by a sense of higher purpose, a destiny, as it were, of which we are only now becoming aware. This new vision synthesizes a century of scientific and theoretic research into the nature of the human species and our ultimate place and role in the evolving universe.

As we have expanded our knowledge of the nature of the universe and the evolution of the human species, we come ever closer to arriving at a scientifically based understanding of our role in the universe. Science's earliest proclamations about the role of man in the cosmos could hardly be considered encouraging in this regard. In the mid-nineteenth century, naturalist Charles Darwin's famous theory of evolution, for all its strengths, portrayed the appearance of all species, including the human one, as a largely random occurrence, dependent on accidents borne out of the statistical vagaries of "natural selection." Darwin and his descendants would consider applying terms such as "destiny" and purpose to the human species and its activities scientifically invalid.[10]

However, recent writings have suggested that the appearance of mankind is not as random and purposeless as Darwin and others have contended. Jesuit anthropologist Teilhard de Chardin, in such works as *The Phenomenon of Man* and *The Vision of the Future*, used Darwinian evolution and biology to speculate that the appearance of humankind is an event critical to the continued evolution of planet Earth. Teilhard went so far as to suggest that humankind itself is evolving into a more advanced form.[11] Biologist Michael Behe, in his book *Darwin's Black Box*, has challenged the notion that the appearance of man in the cosmos is a merely random occurrence.[12] And, by the 1990s, a new field, *Anthropic Cosmology*, pioneered by scientists John Gribben, Martin Rees, and others, provided evidence that the universe's chemical and molecular structure seems so perfectly calibrated to the development of organic life as to suggest that the universe has been "preparing the way" for the appearance of a creature such as man.[13] In their 2000 book, *Rare Earth*, University of Washington scientists Peter Ward and Donald Brownlee took the case for the uniqueness of humankind a step further, convincingly arguing that it is highly improbable, if not impossible, that a life form as complex as the human species exists anywhere else in the universe.[14]

However, for what purpose would the universe summon into existence sapient beings such as the human species? Books claiming to have the "answer" to that question have proliferated in recent years. Such books include Nobelist Christian de Duve's *Vital Dust*,[15] and Frank Tipler's *The Physics of Immortality*.[16] Physicist Eric J. Lerner, in his book *The Big Bang Never Happened*, speculates on the long-term interrelationship of the universe and our species.[17] Michio Kaku, in his recent book *Visions*,[18] and Freeman Dyson, Nobel-winning physicist, in his work *Imagined Worlds*,[19] speculate that the human species' destiny is to eventually develop into a "planetary society" and somehow intervene in the evolution of the universe.

I demonstrate how the convergence of thinking in cosmology, physics, the media, religion, biology, and other disparate areas are leading to the "birth of a new vision of humanity," *The Expansionary Vision of Human Development*. This theory certainly pushes the envelope, proclaiming that the human species indeed has a special role to play in our evolving universe. Toward the end of the book I reveal the startling conclusions being drawn about our ultimate purpose and destiny.

Such ruminations are hardly esoteric or "philosophical." Government and business leaders, if they are to make correct long- and short-term decisions regarding technological development and the economy, must understand the powerful role that the human species will play in the future. Indeed, speculation about such "cosmic" issues is becoming commonplace. In the summer of 1999, in an ABC television series entitled "Brave New World," host Ted Koppel and science reporter Robert Krulwich posed weighty questions about the possibility that the human species does indeed have a special destiny. Popular journals such as *Newsweek*, and *The New York Times Sunday Magazine* regularly run features on related issues, such as the age and size of the universe, human evolution, and the role of the species in the greater scheme of things. An April 2000 issue of *Time* featured a myriad of articles with such intriguing titles as "Will We Keep Evolving?" "Will We Live on Mars?" "Will We Travel to the Stars?" and "Will We Find Another Universe?"[20] Scientific discoveries by Hubble spacecraft and the Mars Pathfinder mission only fuel the debate over the place of man in the cosmos. Moreover, NASA has created an Astrobiology Program to study the origin, evolution, distribution, and purpose of life in the universe.[21]

This new vision provides startling answers to the questions: Why us? Why here? Why now?

A Truly Human Future

We are entering a *human* future, in which the very shape and direction of all aspects of the universe will be deeply influenced by the actions of the human race and its descendants. For example, the proposed terraformation of Mars, the creation of an Earth-like environment on the Red Planet, encompasses more than a planetary facelift—it will represent a thorough "humanization" of that currently lifeless sphere. The significance of humanity's influence on the cosmos will become clearer later as we explore vitalization and the emerging vision of human development.

This future will be one based on human values. It will be a humane future, reflective of our core values—growth, progress, optimism, hope, and altruism. The very act of vitalization, the bringing of life to other worlds, implies that we are acting through exclusively human values—the desire to improve our surroundings, that is, to enrich, embellish, and make the world a better place. While human imagination and energy will build this new world, our values will shape it.

The future is an act of will. It is dependent on the eagerness of humankind to recognize its destiny and its courage to accept the responsibility to take action to achieve its goals. The future is not written in stone—to succeed we will have to exhibit ingenuity, will power, and perseverance. Therefore, regard this book less a prediction of future events than a prescription for how and why to achieve that future. Any work presenting a vision of the future—Marx's books on the economy, or a corporate CEO's mission statement—describes for its readers what they must do to make that future happen. Unless each reader takes action to make the mission a reality, the written prescriptions for change are nothing more than words on a page. Ultimately, the future is in our hands.

The future is also an act of love—of our species, the universe, life, and creation itself. We must rekindle our appreciation for the wonder that is the human species—our accomplishments, our breakthroughs, and our creativity. I long ago became convinced that our species could only achieve anything of worth if we cherished our humanness. You have to love yourself before you can love the world enough to want to change it for the better.

We do well to bear in mind that as we help the world progress, we too will change. By the very act of transforming the cosmos we will become transformed—physically, emotionally, at our very core. We are now about to begin a journey into the future, to be challenged in ways we have never imagined.

Let the adventure begin!

1

Dominionization

Humanity Ventures Forth and Learns Nature's Ways

As the twenty-first century begins, the human species is performing scientific and technological feats of wonder that until recently would have been considered impossible, if not downright magical. Our jumbo jets soar seven miles above the ground; we transmit pictures and sound literally "over the air" via radio, television, and satellite; and we can cure diseases that have plagued our species throughout our history.

And the best is yet to come! As the third millennium unfolds, our species will assume control over matter, reshape the planet, develop new energy sources based on the laser and antimatter, and regulate many of the unseen, elementary physical foundations of the universe, such as electromagnetism and the atom. Throughout the twenty-first century, we will essentially redefine our relationship with nature. I label the sum total of these activities *dominionization*, a force which will, in tandem with other dynamic forces described in the coming chapters, deliver tremendous benefits to humankind, including untold prosperity and a longer, healthier life span.

Ultimately, humankind's progress in the area of dominionization will drive our species to achieve a much greater feat, *vitalization*, the spreading of human intelligence and consciousness first across this planet and eventually throughout the universe. We will infuse the cosmos with humanness, bring order to chaos, beauty to the barren, life to the void.

Dominionization of our home planet prepares us well for the vitalization of the planet. Before we can oversee the development of other planets' topography and climate, we must first learn to perform such operations here on Earth. To rocket from one sphere and travel to the stars we must develop highly sophisticated and powerful forms of energy including nuclear fusion and solar power. In addition, dominionization helps humanity achieve the level of material affluence required to pursue vitalization, an enterprise of Brobdingnagian proportions. Not until we achieve a high level of material abundance can we even reasonably expect to reengineer the universe. Fortunately, humanity is rapidly mastering industrial and agricultural techniques and technologies empowering us to eliminate scarcity and create global affluence.

Dominionization's Apex in the Macroindustrial Era

The achievement of dominionization is occurring in the midst of what I term the *Macroindustrial Era,* a watershed period in societal evolution. In a sense, this era resembles those other special moments in history when the species suddenly "catches fire" and frantically immerses itself in an orgy of creation and invention. Around 3500 BC, the Sumerians, inhabiting a region near the Euphrates River in ancient Mesopotamia, located within the borders of modern Iraq, unexpectedly experienced an explosion of technological innovation. Within a matter of months or even less, the Sumerians invented *wheel transport, metallurgy, sailing ships,* and *wheel-turned oven-baked pottery,* and possibly most notably, writing.[1] During the Renaissance, a nearly somnolent species seemed to wake up and engage in unparalleled creativity. The artistic creations of the fifteenth and sixteenth centuries are unequalled in their beauty and originality—the works of Titian, Leonardo, and Michelangelo speak for themselves. During the same period, humankind discovered and colonized new continents, and simultaneously laid the framework for modern-day science.

As we enter the third millennium, humanity has reached a turning point in its own history: scientific and technological advancements are finally reaching a "critical mass" that will enable our species to play the crucial role the universe requires for its continued development. Before I show how such developments are empowering the species to achieve dominionization, let me first explain why I label the forthcoming period Macroindustrial.

The term *macro* refers to anything that exists on an immense scale or in large quantities. It certainly befits a period in which humanity will expand and extend its many capabilities. We will be erecting mile-high cities—buildings containing hundreds of thousands of people. We will colonize space, travel at supersonic speeds between cities and continents, and extend the human life span by decades if not centuries. The second part of the term, industrial, infers that human activity in this era will be directed primarily toward the production of tangible objects—energy, consumer goods, and new forms of transportation.

I specifically included the term *industrial* to counterpose my view of the future to the currently commonly held belief that we are entering an "Age of Information," an idea Alvin Toffler,[2] John Naisbitt,[3] and others popularized in the last decades of the twentieth century. Proponents of the Information Age concept claim that in the future the main object of human endeavor will be the production, consumption, and transmission of information. In contrast, I contend that in the emerging era the species will direct its efforts primarily to the production of material wealth, goods, objects, and services in a wide variety of areas, including transportation, health, and energy. The "information revolution," of course, will play an important role in the Macroindustrial Era. For instance, innovations in computer and information technology will enhance the ability of scientists, researchers, and the general public to make scientific breakthroughs and invent new products that will accelerate human progress in this new era.

A few general characteristics separate the nascent Macroindustrial Era from all previous human experience. First, in this era humanity will extend its domination in six separate dimensions: *time, space, quantity, quality, size,* and *scope.* I will explain these dimensions forthwith. Second, this era will be truly "global"—every nation and race on the planet will contribute to humankind's mission. Thirdly, the species will finally gain such control over nature that we will be in position to direct the evolution of the universe as well as our own species. In other words, in this era we will be achieving dominionization.

The Many Dimensions of the Macroindustrial Era

For purposes of clarity, when I introduced this concept in the mid-1990s in my book *Seizing the Future,* I dealt with Macroindustrial Era activities that would extend humanity's influence over six separate dimensions.[4]

The first dimension is *time* itself. While we have not discovered how to add hours to the day or months to the year, in this new era we will certainly increase the number of goals that can be achieved within this supposedly fixed resource. We are able to extend the human life span through genetic engineering and other medical miracles. Each individual now has more time to spend making his or her contribution to society's mission. The introduction of superfast transport such as the maglev supertrain (operating at speeds of 250 miles per hour and higher) will enable individuals to spend less time traveling and devote more time to accomplishing other goals. We also will shrink the time it takes for us to accomplish tasks. Soon robots will perform household chores and begin to perform more menial jobs such as hamburger flipper and hospital orderly. Humans will be liberated, able to spend their time performing higher-level tasks. Computers will continue to obliterate the time it takes to perform scientific research and manufacture goods.

The next major change in the Macroindustrial Era is humanity's increased control over both the inner and outer reaches of physical *space*. Throughout the early twenty-first century and into the next millennium we will be exploring more of the universe. We will return to the Moon and eventually settle Mars. The International Space Station portends greater cooperation in this effort on the part of Russia, the United States, and several other countries. We are beginning to conquer space beneath the Earth's surface. Japan, for instance, plans to build underground cities containing homes, offices, and power plants. Extensive underwater development will also become reality— on ocean platforms, artificial islands, and structures attached to seamounts. The species will soon control the inner space of matter itself. We are developing a new science, nanotechnology, in which we will use the very atom itself to cheaply and simply build any material we so desire—spaceships, houses, even new skin.[5]

The next dimension humanity will master in the Macroindustrial Era is *quantity*. We will apply novel technologies to create an abundance of food and products that will help us eliminate scarcity and poverty. Breakthroughs in manufacturing and the development of new forms of energy will enable us to bring the billions of people on the planet up to the standard of living of those in more developed nations. Robots, computers, and the advent of the "cybernetic factory" will enable us to produce a higher quantity of goods. As we will see later in the chapter, our achievement of dominionization over energy sources through new developments such as nuclear fusion power, advanced forms of solar energy generation, and nonpolluting hybrid fuel-cell engines will help us produce food, homes, and manufactured goods in greater capacity.

We will also develop technologies that will help us create products and services of a higher *quality*. Researchers will apply biotechnological techniques to develop more nutritious foods, such as the Calgene tomato. Macroengineering and new discoveries in material science will enable us to produce customized high-quality goods made out of "smart materials" that are incredibly durable and even able to adapt to outside temperatures. A better-educated and highly skilled public will facilitate our mastery of the *quality* dimension. Improvements in the way information is delivered, such as the Internet and the CD-ROM, will make it easier for all of us to contribute to the innovation and invention process. Soon, everyone who so desires can be a scientist, inventor, and even an armchair "natural scientist" contributing thoughts and insights to the scientific enterprise.

The Macroindustrial Era will also witness the expansion of the *scope* of both production and consumption to global proportions. No longer will only the Western and Northern Hemispheres enjoy the benefits of abundance and affluence. If trends continue, people in developing countries will participate in the global economy as producers, workers, and consumers. In the twenty-first century, the populations of countries such as India, Brazil, China, and Indonesia will represent an ever-growing proportion of the world's total population. Citizens of these emerging countries are quickly acquiring the skills needed to master fields, such as biology, physics, and computer science, that the species needs to achieve dominionization over this planet. They will exchange these skills on the world market for money and goods and increasingly be able to claim their "piece of the pie." As a result of this process, we are already seeing a rapidly growing middle class throughout Southeast Asia and in parts of South America.

In this era, the human species will fundamentally redefine the concept of *size*. The sheer immensity of such projects inspires awe. Ever-grander skyscrapers—the Houston Tower, 1.3 miles high, and 2500-foot World Trade Center in Chicago, at 210 stories high—suggest that we have nowhere to go but up. Mammoth projects of all types are typical of this era. Japan and Monaco are building artificial islands in the middle of oceans and lakes, which can accommodate large numbers of inhabitants. China is erecting the huge Three Gorges Dam on the Yangtze River, the largest engineering project on the face of the earth. China is literally reshaping the earth in order to allow cargo ships and passenger liners to sail 1500 miles from the Pacific into China.[6]

The Coming Age of Superabundance

Our efforts to achieve dominionization over the basic forces of nature and over the Earth itself have a major beneficial by-product—the creation of a superabundance of food, goods, sources of energy, and manufactured products. This end of scarcity on a global level is a landmark event in human history. During the Macroindustrial Era, we are redefining the concept of "the good life." More important, the rapid diffusion of wealth and wealth-generating technologies and knowledge is in turn enabling the global population to participate in the dominionization process.

In the agricultural domain, breakthroughs in biotechnology and genetic engineering will deliver to humanity a veritable cornucopia of new agricultural products that resist disease, frost, and infestation, and have a longer shelf life. Cell factories and plant tissue technology are making possible the mass production of vegetable and fruit in artificial environments. Hydroponic plants will grow in waterless soil! The sum total of these efforts will provide goods and food to the multibillions inhabiting our planet—for the first time in human history the world's population will be well fed, well clothed, and comfortably housed. And a very large population could be served, perhaps 40 to 50 billion people or more.

Companies in the United States have been at the forefront of the global effort to apply genetics to the enhancement of the national and international food stock. The U.S. Department of Agriculture takes this process very seriously and is trying to assist such companies. In 1999, the USDA announced that it would set up a national gene research center at Cornell University. According to Judy St. John, an associate deputy administrator with the USDA's Agricultural Research Service, "The USDA-funded center will aid researchers around the country and the world in the quest to discover all the genes in grains—like corn, wheat and rice—and plants in the family that includes tomatoes, potatoes, and peppers."[7] The information generated at this new center will facilitate the development of better food products. Once scientists identify a gene's function, they can experiment to determine whether that gene can be rebuilt in a way to make it more effective. For instance, if we find genes that are disease-resistant, we can simply move them into plants that lack such resistance. This could actually help the environment immeasurably, since such enhanced plants would require fewer chemical pesticides.

A striking example of how our efforts to achieve dominionization have helped us generate abundance in the agricultural domain is the creation of the so-called "supertree." Companies such as International Paper and West-vaco are currently genetically engineering trees that are strong, disease-resistant, and can grow just about anywhere, even in desolate fields. Such trees will grow at such accelerated rates that they will reach 100 feet in less than 10 years. Biotechnologists are trying to isolate and manipulate genes to ensure that such trees will serve as the source for forest products such as paper. This process serves as a solid example of how humankind is vitalizing the earth itself. This supertree is actually a wholly human product—we are not just replicating nature; we are adding human intelligence and creativity to the Earth's biosphere. Human consciousness, not *nature*, has brought the supertree into existence.[8]

Our efforts to learn nature's secrets and reshape the planet have dramatically upgraded the global standard of humanity, even as the world's population steadily expands. As we enter the twenty-first century, the material status of the world's peoples is improving dramatically from where it was even 30 years ago. To quote a recent newspaper column assessing the state of the world at the beginning of the new millennium, "the evidence indicates that things are getting better for almost everyone."[9] All barometers of personal and economic well-being are improving, even in the developing world, including infant mortality, life expectancy, literacy, and caloric intake.

The writer goes on to say that "food production per capita continues to soar while food prices decline."[10] This optimism regarding the ability to feed the world was confirmed in late 1999 by a communiqué from the Paris-based International Federation of Agricultural Producers. Even as the UN population clock ticked off the 6 billionth human born in mid-October of that year, the Federation affirmed that grain was as plentiful and inexpensive as ever. Food industry experts proclaimed that the world's rapidly expanding technological know-how ensures that food production will greatly outpace demand. Furthermore, the report goes on to say that this situation can only improve as we tap the enormous food-growing potential of countries such as Kazakhstan and China. The People's Republic of China celebrated its fiftieth anniversary with the announcement that, after four years of bumper crops, the country could not only feed itself but would begin to export certain food products. The Federation declared that genetically modified food technology would benefit poor countries' farmers, since the crops produced through this technology would be resistant to pests, disease, and drought.[11]

Humankind's continued dedication to scientific research and its willingness to aggressively adopt new technologies will ensure that this "age of superabundance" continues.

Technology Can Solve Its Own Problems

Humankind's invention of the "supertree" illustrates one of the least-acknowledged benefits of the *hyperprogress* occurring during the Macroindustrial Era: although economic, technological, and industrial growth occasionally cause problems, such as pollution and possible reduction in the supply of some natural resources, our technology ultimately generates solutions to the very problems it creates. Our civilization requires wood and paper products to continue progressing. In the process we temporarily reduce the available source of these wood products, namely trees. However, our resourceful species just as quickly replaces these commodities—in this case we applied genetic engineering techniques to produce greater quantities of wood.

In fact, economic growth and technology directly counteract environmental degradation. Studies indicate that while a developing country's early economic growth initially can lead to pollution and waste, once that nation achieves true prosperity it then possesses the resources to clean its air and purify its water. In his 1999 book, *Earth Odyssey*, environmentalist David Hertsgaard reports that the poorest cities, not the most prosperous, were usually the most polluted. The citizens in these places can buy cars, but cannot afford cars with catalytic converters. According to the World Bank, once a nation's per capita income rises to about $4000 in 1993 dollars, it produces less of many pollutants per capita. At this income level a nation can now afford to purchase the technology to purify its coal exhausts and the sewage systems that treat and eliminate a variety of wastes.[12] Although China is switching to cleaner technologies such as nuclear, it still favors the use of its locally abundant, and therefore cheaper, resource, coal. We can predict that once countries like China become more affluent they will have the wherewithal to clean up their atmosphere.

Technology is now being used to deal with waste produced by tanker accidents and other unexpected events that can send millions of gallons of oil or other chemicals gushing into our pristine lakes and oceans. One novel method, *bioremediation*, uses microbes, bacteria as it were, as a veritable cleanup crew, for everything from nuclear waste to oil spills. This new environmental technology is based upon the notion that bacteria are the perfect agents to literally "eat" industrial waste. U.S. Microbics is one company that specializes in this increasingly popular technique. (Publicly held, its stock symbol is BUGS . . . seriously.) In early 1999, it announced that its new production plant had commenced shipping microbial blends to treat hydrocarbon-contaminated soil. U.S. Microbics used biotechnology, bacteria mostly, to clean up sewage wastes, including diesel oil spills. According to the company, "naturally occurring bacteria blends are used to convert the hazardous diesel fuel into harmless, earth-friendly, chemical byproducts."[13]

A larger challenge lies in the application of bioremediation to nuclear wastes. Scientists have discovered a radiation-tolerant bacterium called *D. radiodurans.* This natural anomaly has a number of unique abilities, including an unprecedented capacity to repair human genetic damage. The U.S. Department of Energy thinks that the microbe, after appropriate genetic manipulation, might help detoxify the thousands of toxic waste sites that include radioactive materials. Early studies show that *D. radiodurans* has successfully degraded an organic toxin common to such waste sites.[14] Bioremediation might actually be the ultimate solution to the thorny problem of nuclear waste disposal.

As mentioned, progress does engender some unfavorable by-products. However, it also spawns the science necessary to both solve these problems and generate further progress. In the next sections you will encounter some of the Herculean efforts to achieve dominionization, including our efforts to tame the atom and control the weather.

Dominionization over Inner Space: The Promise of Nanotechnology

Through dominionization, humankind is gaining control over the elementary dynamics of nature. Nowhere is this better illustrated than in our attempts to control the behavior of the atom itself. Earlier in the twentieth century, American scientists working on the Manhattan Project in Los Alamos, New Mexico, successfully split the atom, an accomplishment that led to the creation of the atomic bomb. It also opened the door to the development of nuclear energy, which provides a significant proportion of the electricity generated throughout the world.

While we can split the atom and use it for energy and defense purposes, we have not yet achieved a true command over this essential building block of matter. Until recently, the inner workings of the atom have been considered an unpredictable frenetic maelstrom of activity outside the realm of human control. Now, a burgeoning new science known as *nanotechnology* tantalizes the species with the possibility that we might indeed corral this elementary unit and direct its behavior. In theory, nanotechnology would equip our species with the ability to construct material—clothing, food, body parts—from the bottom up, one atom at a time. In essence, we would suddenly possess the power to "grow" almost any object—a tool, a human arm, or a rocket ship—by simply selecting the correct atoms and then programming nanocomputers to construct the object.

If we do succeed in manipulating matter to this degree, scarcity and poverty will become faint memories from a bygone era. *Nanomachines* about the size of viruses could take a pot full of ingredients, and build an automobile one atom or molecule at a time, simply by placing those atoms and molecules just where they ought to go. Ben Bova, in his book *Immortality*, described the process thus: "Building an auto would be like a swarm of invisible genies working more silently and swiftly than a noisy, clanking factory." From a heap of charcoal dust, ordinary carbon soot, nanomachines would produce an automobile, an airplane, or any other object. Imagine for a moment a car with the structural strength and lightness of diamond. Nanotechnology will enable us to develop "smart materials" that could adjust to changes in the environment and advanced materials so durable that they could withstand the stress of protracted space travel.[15]

Ed Regis, in the volume Nano (aptly subtitled *Remaking the World—Molecule by Molecule*), describes how this process might work. A fleet of ultratiny invisible robots called *nanoassemblers* would break down the chemical bonds of the molecular ingredients such as grass and water, and then reassemble the molecules of these ingredients, such as hydrogen, carbon, and nitrogen.[16] According to Xerox researcher Ralph Merkel, these diligent, tireless microrobots could be programmed to recombine these basic building blocks of matter and build a wide range of useful products, like food or a piece of furniture. These universal assemblers would use their submicroscopic legs to put these molecules together like Legos.[17]

K. Eric Drexler, a noted theorist in the field, thinks we can build such a nanoassembler by taking what he labels an "evolutionary" approach. First, engineers would build the smallest machines our technology will allow. These machines would build still smaller units, which in turn would build tinier devices, until a nanometer-sized *replicator* has been perfected. As its name suggests, this replicator would be designed primarily to build working nanomachines.

A dedicated cadre of many individuals and organizations are feverishly involved in bringing this new science to life. Research on nanotechnology is being performed in laboratories, universities, and organizations in the United States, Japan, and Britain, at places such as MIT, Stanford, Cornell, the University of Michigan, NASA, IBM, Lucent Technologies (Bell Labs), Xerox, SRI International, and other scientific institutions. NASA is looking at computer models of self-maintaining materials for building the spacecraft of the future. I recently learned that U.S. Army strategic planners are seriously studying the possible military and industrial applications of nanotechnology. Governments around the world are already pumping millions into nano-technology research.[18] The Japanese government, through its Ministry of International Trade and Industry (MITI), is investing in companies delving into the mysteries of the ultratiny. In 1998, the University of Washington opened its Center for Nanotechnology in the city of Seattle. The control of the atom will require a great deal of partnering among organizations of disparate backgrounds. For instance, The University of Toronto and Energenius Inc. in 1998 announced the opening of the Energenius Centre for Advanced Nanotechnology. ECAN is dedicated to training advanced students in the area of semiconductor nanotechnology for future device development. It is working on projects with a number of organizations, including the National Research Council of Canada, the Cornell Nanofabrication Center and the University of North Carolina at Chapel Hill's Nanomanipulator Project. In 1998, the German government announced that it would create five nano-technology centers.[19]

How close are we to transforming these concepts into reality? This science has taken its first important baby steps. In 1989, researchers at IBM used an instrument known as a scanning probe microscope (SPM) to arrange 35 xenon atoms on a nickel surface to form the letters of their company, IBM. To construct this atomic-sized company logo, the SPM dragged a canti-levered arm across a surface to move atoms into play. By getting these atoms to sit still long enough to spell out IBM's name, scientists demonstrated that we could marshal the forces of individual atoms to create useful objects. Chemist Rick Smalley, director of Rice University's Center for Nanoscale Science and Technology won the 1996 Nobel Prize for the discovery of a new carbon molecule called *fullerene*, also known as "buckyballs." Smalley predicts that eventually this technology will empower us to use our fingers, extensions of our own body, to reach down and touch an atom, a molecule, or an enzyme, and literally move it from one place to another! Moreover, our accuracy in manipulating the ultratiny is improving yearly. Already, ful-lerene has been used to create a nanotube, a cylinder with open or closed ends that NASA computer simulations predict could actually work like a gear if built. In 1996, another team of IBM physicists in Zurich built a nanoscale abacus, with each bead of the primitive accounting tool consisting of a carbon molecule.[20]

A recent *Newsweek* piece exclaimed that "playing games in the nanoworld holds the Promethean promise of giving society any material it wants at virtually no cost."[21] The outcomes the article describes could someday be real: 500-story skyscrapers made from diamond instead of steel, paint that changes color, fabulously powerful microprocessors, spaceships that have the strength of titanium but the weight of plastic, and medical robots that can kill cancers and viruses in our bloodstream. NASA is looking at computer models of self-maintaining materials for building the spacecraft of the future.[22]

Much of the energy, encouragement, and direction needed to achieve this Holy Grail of physics comes from members of what I label the "expansionary vanguard." Many of these people working to decode the mysteries of the atom toil outside the parameters of "establishment" science. A true maverick, Richard Feynman, the legendary physicist, was the first to suggest the underlying idea that evolved into the science of nanotechnology. In 1959, Feynman ended a speech at Cal-Tech, called "There's Plenty of Room at the Bottom," with an intriguing statement: "I am not afraid to consider the final question . . . can (we) arrange atoms the way we want, all the way down."[23] In 1981, the aforementioned K. Eric Drexler, then a 26-year-old graduate student, published a paper in conference proceedings describing the operational principles for a "science of the ultratiny" envisioned by Feynman over 20 years earlier. He named this new science nanotechnology. He formed the Foresight Institute in California, which today hosts numerous conferences on this new science and serves as a clearinghouse for information on the subject.[24]

Such individuals and organizations serve as catalysts for breakthroughs that will lead to a successful dominionization of our physical world. And judging by the proliferation of Web sites dedicated to nanotechnology issues, the number of researchers, supporters, and enthusiasts will only increase. Computer simulations and other novel technologies are making the research and discovery process more accessible to the layperson and amateur scientist. Who knows where the major breakthroughs in this exciting new field will emanate from?

Bill Spence, editor of *NanoTechnology* magazine, predicts that average citizens will have micromanufacturing sites right in their homes. People will build their own wristwatches and computers as easily as they could print out an article off the Internet. Certainly, nanotechnology will aid our species as humans migrate throughout the galaxies. As we become more physically distant from each other, we will have to develop ways to become self-sufficient. Nanotechnology, once perfected, will enable small groups of space explorers and colonizers, far from centralized production facilities, to manufacture anything they need on the spot with available non-user-friendly materials that exist on their current asteroid or planet.

Nanotechnology brings humankind into the very center of the act of creation. True, we are not manufacturing matter out of "nothing." However, we are creating form from the amorphous, order from chaos, direction from the random. Nanotechnology will allow us to fabricate anything as long as we have at hand its basic constituent molecules. Each individual will become an alchemist of matter, a veritable Zeus commanding molecules to dance and calling atoms to formation. This new power will change us all. "When you start thinking what you can do if you could really position atoms wherever you dream of them, it looks like a wonderful future," says Rick Smalley.[25] If we become rulers of the molecular world, how soon will we become masters of the universe?

In early 2000, the U.S. government took actions that reflected this growing awareness that nanotechnology can fundamentally change the relationship between humans and nature for the better. In early 2000, President Bill Clinton proposed that the government spend almost a half-billion dollars on research and development in the field of nanotechnology.[26]

Dominionization's Energy

It is safe to say that our species has been successful in generating the energy it needs to fuel its *Macroindustrial Society*. Up to this point, our species has largely depended on fuels from the earth—coal, oil, wood—to run our machines and heat our homes. The human species took a significant evolutionary step forward when it learned how to transform these innate objects into sources of energy to power the turbines that fuel our entire industrial enterprise.

Unfortunately, as we enter the new millennium, we are gradually depleting the supply of such energy sources. Under normal economic conditions, fixed resources such as oil and coal become scarcer over time. However, newly prosperous Asian and South American nations competing with the developed countries for the available supply of natural resources put even greater strains on the world's oil supply. In 1997, the International Energy Agency (IEA), a Paris-based think tank, predicted that by 2010 the global demand for oil would rise by 30 percent, and Asia would be the leading customer for petroleum.[27]

Eventually, the world will need more petroleum than we can possibly refine. In articles in *Scientific American* and *Science*, oil industry geologist Colin Campbell predicts that by 2010 we may be confronting an "oil peak," when worldwide demand for oil exceeds supply. There are several reasons for this coming disjuncture. First, world demand for electric power is rapidly expanding. Even the continued economic problems in Asia and Russia throughout 1999 could not dampen this demand sufficiently to change Campbell's prognostication. Global demand for energy is almost guaranteed to grow dramatically, some think by as much as 60 percent by 2020. Second, supply of petroleum is stagnating. For instance, in the 1980s, the world produced 220 billion barrels of crude, but since then oil companies have found only 91 billion in new discoveries to replace that supply. Scientists with the U.S. Geological Survey (USGS) warn that this production peak could come as early as 2005. University of Colorado researchers predict such a peak in 2020.[28] At a recent meeting of the World Energy Council, even oil company executives now quietly admit that we can only produce sufficient levels of "light sweet crude," the oil that is relatively easy to pump and refine, for at best another 40 to 50 years.[29]

Events in 1999 and 2000 revealed just how fragile and unpredictable is a world economy so dependent on oil as its main energy source. In early 1999, OPEC countries mustered the self-discipline to impose an extremely effective quota on petroleum production. Within a few months of OPEC's restriction in oil production, the price of crude oil began to increase precipitously, jumping from about $14 per barrel to over $24 in less than three months. In the United States, gas prices at the pump increased to levels consumers had not experienced for years. In September 1999, OPEC met again, in Vienna, and stated unequivocally that it would restrain growth in the supply of oil, regardless of demand and price, throughout 2000. Charles Schumer, U.S. senator from New York, called on U.S. Energy Secretary Bill Richardson to release crude from the country's Strategic Petroleum Reserve to force a decrease in world oil prices. "The United States can and should defend itself from what amounts to nothing less than economic warfare," Schumer wrote in a strongly worded letter to the energy secretary.[30]

He would no doubt be disappointed in the response of the Executive

Branch, especially its boss, the President, to the energy issue. By 1999, Clinton, under the sway of a new book called *Natural Capitalism*, had adopted, and regularly shared with leaders of developing countries, the idea that developing nations do not need to be "energy hogs" to become rich. According to Clinton, using large amounts of energy, including oil, to fuel a nation's economy, was the old-style pattern of the industrial revolution. In other words, to solve energy shortfalls and concomitant higher oil prices, the world should curtail its demand for energy, not increase the supply.[31] One is left to conclude that a government that considers energy production and consumption an inconsequential factor in economic growth would hardly take steps to reduce worldwide energy prices, especially such actions as radical as Senator Schumer's.

As the second millennium came to an end, Clinton's theory about the relative unimportance of energy consumption in economic growth was sorely tested. The Asian economies, including that of China, were rebounding from their 1997 financial crises. These countries, evidently following what Clinton derisively labels old-style strategies for growth, gradually increased their petroleum consumption. The world was now competing for increasing amounts of a commodity already made scarce by the OPEC cartel's stingy production schedule. The price of oil continued to rise, and by November 1999 inflation worldwide and in the United States started to move upward. The Federal Reserve hoped to control further inflation by raising interest rates, but this move began to rock Wall Street—the Dow Jones Industrial Average dropped in response to what analysts considered an economy-dampening strategy. Throughout the first half of 2000, the Fed's continued to raise interest rates, causing increase jitters on Wall Street that essentially ended the bull market of the late 1990s.

As the third millennium begins, the energy outlook is hardly improving. The International Energy Agency projected a sharp acceleration in worldwide demand for petroleum and only a marginal increase in supply. The IEA reported that "A first look at world oil demand in the year 2000 shows a very different picture from the last two years, with the demand focus returning to non-OECD countries." It projected that Asian countries, especially China, would put enormous strains on the world's petroleum supply.[32] (A few months later India bore out the IEA projections about world economic recovery, claiming a one-year jump in GNP of over 8 percent.[33]) In 2000, to make matters worse, an unexpectedly frigid winter sent home-heating oil costs skyrocketing throughout much of Europe and the United States. By May 2000, analysts were warning that increased fuel prices would hit the

U.S. manufacturing and transportation sectors particularly hard. Companies such as Procter and Gamble saw their profits squeezed by rising prices, and a number of airlines were forced to raise airfares to pay for the rising cost of fuel.[34]

The only way the species can transcend its dependence on petroleum is to discover substitutes for this limited commodity. Fortunately, our species' dominionization implies the development of more complex and sophisticated forms of energy than the burning of oil, natural gas, and coal, a fairly primitive process. We will manipulate the atom, as in nuclear fission or fusion power, or tap the power of the sun, to produce ever more abundant and powerful forms of energy. We may even begin to incorporate lasers to power our spaceships and cities from afar.

We must develop increasingly imaginative methods for producing greater amounts of energy to support an expanding global prosperity! Our new energy forms must be sufficiently plentiful, accessible, and cheap to power high-speed rail systems, space missions, and highly productive factories serving the billions worldwide. Moreover, they must be user-friendly, so that space travelers can use and produce them as they explore and colonize distant spheres.

The Power of the Atom

By its very definition, dominionization connotes humanity probing into the deepest recesses of nature, unlocking its secrets and liberating its potential. Certainly, opening up the atom to reveal its power and then channeling that power into the production of energy for mankind epitomizes the dominionization process.

And our energy needs are ever growing. The U.S. Energy Information Administration expects worldwide electricity consumption to double in 20 years. This rapid growth in demand will require that we construct 5000 new 300-megawatt power plants. World electrical energy consumption is increasing at about 2.5 percent per year, even though currently more than 40 percent of the earth's 5 billion people are not connected to electric power grids. According to the U.S. Department of Energy, by 2015 the consumption of electricity in the so-called developing countries—about 80 percent of the world's population—will reach parity with the consumption levels in the industrialized sector.[35]

In the late 1990s, nuclear power was among the more popular energies of choice. Countries such as China, Taiwan, Sweden, and Japan were moving toward nuclear power to meet their short-term energy needs (and minimize their dependency on petroleum). While in the United States nuclear power production is remaining stagnant at 20 percent of the total electricity market, Japan and South Korea predict that nuclear energy will supply 40 to 50 percent of their power needs in the next century. Japan is developing its own "fast breeder" reactor. This type of reactor is considered a highly efficient generator of nuclear energy, because it recycles its own spent fuel.[36] Germany receives about 30 percent of its electricity from 19 atomic power plants. France's electrical network is 90 percent dependent on nuclear energy.

For all the negative publicity nuclear power receives in the United States, it is noteworthy that American companies such as General Electric are in the forefront of the international nuclear power station business. U.S. nuclear technology companies have invested billions of dollars in erecting power plants inside Japan, Taiwan, South Korea, North Korea, and China. These companies are also eyeing as potential customers Turkey, Hungary, and Indonesia. Westinghouse anticipates that the Chinese nuclear market will boost its profits. Companies from the Western countries are planning to help China build at least 10 new large nuclear plants by the turn of the century. China will be spending up to $100 billion on nuclear energy over the short run.[37]

In the United States, nuclear power has been mired in a political and regulatory quagmire. Since the Three-Mile Island power plant accident in 1979, no new nuclear power plants have been approved for construction in the United States. Regulatory agency red tape and environmental pressure groups' political tactics have made it prohibitive for companies to bring nuclear power plants online in the United States. Not unexpectedly, U.S. dependence on foreign oil continues to increase. The German nuclear power industry is also coming under fire. The left-leaning government of Gerhard Schroeder, feeling pressure from the Green Party, even tried to shut down some of Germany's nuclear power plants, but faced stiff resistance from industry as well as scientific organizations.[38]

The full effects of cutbacks in nuclear power production in the United States were not felt in the late 1990s mainly because petroleum remained cheap. By 2000, however, as world prosperity returned with a vengeance, global energy demand began to increase. This cavalier attitude toward nuclear power and energy exhibited by the United States, Germany, and other developed countries has unwittingly put these nations at the mercy of the international oil market. The price increases of petroleum-based products experienced by these countries is due in part to their refusal to expand nonpetroleum energy sources such as nuclear fission.

The Power of the Stars

Nuclear fusion, or simply fusion, is a form of atomic power. It is the opposite of the process described in the last section, *nuclear fission*, the type of fuel most people think of when they hear the term "atomic power." In fission, we obtain energy by *splitting* the atom. *Fusion* produces energy as a result of very light atoms *colliding* and *fusing*. This collision and subsequent fusing yields heat energy that can be used to make electricity. This is the same process by which our sun and other stars produce heat energy.

Fusion power, when it finally becomes commercially feasible, holds the promise of dramatically improving the global standard of living. Since the source of this power is hydrogen, we could conceivably use any substance that contains hydrogen, such as water, to power an entire civilization. According to Dr. David Baldwin, Senior Vice President of General Atomics Fusion Group, a major fusion energy research organization, this new form of energy is an ideal energy source for future generations. Its benefits are that it is "safe, inexhaustible, environmentally benign and . . . a fuel source that is readily available to all nations."[39] And it is probably available throughout the solar system and beyond.

While the process of generating fusion-based energy seems simple in theory, mastering the physics of nuclear fusion has challenged even our most brilliant scientific minds. Happily, we are finally making progress in establishing a viable form of fusion power. In congressional testimony on energy security, Baldwin stated that "since 1980 the fusion power produced in experiments has increased a million-fold with the larger fusion facilities now producing over 10 megawatts for several seconds—in electrical equivalent, enough to power a town of a thousand." Some individual tests have exceeded those levels.

In 1994, the Tokamak Fusion Test Reactor (TFTR) in Princeton, New Jersey, set the world record for fusion power by generating 10.7 million watts for about 1 second. This amount of energy could supply power to about 10,000 homes. In late 1997, another major fusion program, the Joint European Torus (JET) project near Oxford, England, produced 12 million watts of power, a world record at the time. The JET record output was six times greater than the amount of energy any controlled fusion experiment had produced just five years earlier. Many feel that a commercially viable fusion power industry is not that far in the future.[40]

Baldwin does not underestimate the difficulties we face in developing nuclear fusion. Re-creating the energy process of the sun and stars here on earth and harnessing its energy, he said, is "the most difficult scientific and technological task ever undertaken." However, he felt that over the last decade a significant number of theoretical technical challenges have been overcome. So why do we not have a viable commercial fusion reactor in 2000? As Baldwin testified, "For several years, the U.S. fusion community has been ready to take this next step and has proposed domestic facilities to accomplish it, only to have them fail for lack of funding."

The funding of fusion research has become a bit of a mystery. In the mid-1990s, the U.S. Congress had cut funding for the Princeton TFTR just as it was achieving success, a move that led to the installation's closing. In 1996, the United States also eliminated funding for the International Thermonuclear Experimental Reactor (ITER), which was regarded as the final step before design of commercial fusion reactors. The European Union, Russia, and Japan eventually reduced their funding after it became clear that the United States was eliminating its support. America still supports fusion research to the tune of at least $700 million, but that support is directed at defense projects and pure research. When a commercially viable fusion reactor comes online, it may not be an American invention.[41]

By 1998, it appeared that ITER was dead. Many scientists had abandoned the project after the United States withdrew funding. At a recent meeting in Madison, Wisconsin, magnetic fusion researchers discussed the prospects for a cheaper version called ITER Lite, now being studied by the international partners. The researchers also considered an alternative approach based on a series of smaller machines that could be hosted by several nations and called for a sweeping review of magnetic fusion science. (Curiously, representatives of the Japanese fusion program attending the conference implied that their country would not participate in ITER Lite but would go it alone in developing advanced fusion reactor technology.)[42]

Signs abound that less developed countries recognize the potential value of this nascent source of nuclear fusion and are prepared to pay good money to become players in this newest energy game. In July 1999, Canada announced that it was considering selling its entire nuclear fusion program, including its storehouse of research material as well as its hard technology, to Iran. Although selling the program made economic sense, by late 1999, Canada announced that it was backing down from the idea—officials in Ottawa said it had been advised by U.S. officials that Tehran might be able to use the technology from the fusion experiments to develop atomic weapons.[43]

In spite of these technological setbacks and political machinations, it is almost certain that a commercial reactor will be operating in the relatively near future. We have already made major theoretic and experimental breakthroughs, and with increased funding the rest of the scientific challenges will be overcome. Funding will increase as traditional energy sources, like fossil fuels, grow ever scarcer. At that point, nuclear fusion will become a major power source of the Macroindustrial Era. It will power our high-speed rail systems and heat mile-high cities. And most important, for the next several centuries it will most likely serve as the fuel for space travel, thereby facilitating our vitalization mission.

Power from Space and the Sun

Space exploration and colonization *will* play a major part in humanity's establishing of dominion over the energy process. The public would be quite surprised by the elaborate plans being drawn to have the space venture play a strategic role in the quest for new sources of abundant energy.

For instance, the space mission will play a major role in humankind's attempt to perfect the nuclear fusion process. According to experts, nuclear fusion requires large amounts of two elements, deuterium and helium-3, to operate at peak efficiency. We can extract one of those elements, deuterium, from water at modest cost. However, the other element, the light gas helium-3, is relatively rare on Earth. Experts such as Gerry Kulinski and his colleagues at the University of Wisconsin claim that the lunar surface contains an abundance of helium-3, implanted there over the eons by the solar wind. Our choice is simple: We must go to the moon to obtain helium-3 in the quantities we need for efficient nuclear fusion reactions.

Dominionization is rarely easy. Initially, the species will incur a huge cost extracting helium-3. We must ship to the moon the mass of equipment needed to extract helium-3 from the lunar surface, establish a mining operation, and then transfer it back to Earth. But these mining operations should become cost effective and eventually extremely profitable. The moon contains sufficient amounts of helium-3 to provide the species several centuries of fusion-based energy. Moreover, as we establish bases on Mars and other spheres, the moon will serve as a source of energy for the new astrocolonies.

To achieve grandiose visions such as vitalization we must think in broad *expansionary* terms. Certainly, a moon base is ambitious. Equally grandiose are the plans currently being formulated for a space-based solar power system. This system would require that we launch solar power satellites (SPSs) into geosynchronous orbit above a fixed point on Earth's equator. Perpetually facing the sun, each solar satellite would collect power from the sun in its extremely large photovoltaic cells. Each SPS would send microwave power beams to a flat array of antennas on Earth covering several square miles of ground. The base on the ground would convert these beams into electrical current, and then feed energy into the neighboring power grid. In some designs, the system would require 60 satellites, each about 6 miles in length and 4 miles in width. Each SPS would be able to supply the power needs of a mega-city such as New York or Chicago.

In the late 1990s, NASA began looking seriously at the solar power question. The study team consisted of participants from NASA, universities, government agencies, and industry. It concluded that for space solar power systems to be cost effective, they must cost no more than $1 to $10 billion to begin generating power commercially. They must serve global markets, be modular and easily constructed, and be capable of launching on space transportation systems that are "common to other markets, not unique to space solar power."[44]

The Princeton-based Space Studies Institute is one of many other groups looking into the possibility of providing solar energy from space. According to the institute, we should consider the moon a site for the construction and launching of the SPSs. The moon offers a low-gravity environment that would facilitate the manufacturing of SPS parts and make an SPS launching more energy-efficient. Of course, we will have to build and maintain lunar work bases. Such bases can be more cost efficient if they can house mining and construction crews.[45] Japan has been actively involved in space-based solar experiments. Moreover, studies in this field continue. The International Space University and the American Institute of Aeronautics and Astronautics have held international conferences to discuss solar power satellites.

The Power of Light

Much of the progress made by our species occurs when people think "outside of the box." This maxim is especially relevant to the many innovations in the energy field.

For the last few centuries, the vast majority of scientists approach energy from the point of view of combustion. That is, how can we "explode" or burn something—wood, uranium, the atom, oil—in order to generate energy. When trying to improve energy efficiency for spacecraft, inventors follow the same model. How do we place a more efficient and more powerful rocket fuel in the rocket? This model's shortcoming is that invariably the fuel on board accounts for the lion's share of the weight of the craft, somewhat inhibiting the launchcraft's ability to get off the ground. In essence, the fuel must lift not only the satellite or payload, but also provide thrust to get itself and its container off the ground. Conventional rockets are weighed down with fuel that accounts for up to 90 percent of the rocket's total weight.

A new "outside the box" concept is on the horizon: leave the fuel source on the ground! In test flights over White Sands Missile Range in New Mexico, inventor-engineer Leik Myrabo of Rensselaer Polytechnic Institute is experimenting with tiny vehicles called *Lightcraft* fired into the sky by power beams of light emitted from a laser.

While the development of the Lightcraft is only in its infancy, Myrabo speculates that eventually such rockets might be converted into minispace telescopes, for example, or perhaps microsatellites that relay phone conversations to any point on the globe. Eventually, he says, large Lightcraft will replace the airplane as the way to travel. "Lightcraft could take 12 people anywhere they want to go in the world in 45 minutes, or directly to the moon in 5½ hours—for the price of an airline ticket!" Moreover, he sees a time when this technology will enable people to fly on highways of light beamed from solar-powered transmitters in space.[46]

Lightcraft's source of power, the laser (short for light amplification by stimulated emission of radiation), itself typifies the bursts of imagination our species often displays as we enter the next millennium. According to Rick Garcia of the Air Force Research Lab at Kirtland Air Force Base, New Mexico, the best way to understand the laser is to think of it as enhanced light. How intense is this enhanced light? According to Garcia, "If one touches a 100-watt bulb, you burn your hand. But when we concentrate light energy into a focused beam of just 10 watts, we can cut tissue."[47]

Myrabo's current experimental model is small by design, only about 4 inches in diameter and weighing less than an ounce. Lightcraft is propelled using a totally novel energy system, which may be the energy paradigm of the future. Lightcraft sits on top of a 6-inch pole. Myrabo fires a jet of air at the vehicle's rim, which causes the vehicle to spin and stabilize before lift off. Then Myrabo works his magic—he beams repetitive pulses of laser light into the craft's cone-shaped nozzle, heating the air in the engine to 54,000°F. This superheated air expands to send Lightcraft skyward.

Myrabo is performing these experiments in collaboration with senior scientist Franklin Mead and other scientists of the Air Force Research Laboratory's Propulsion Directorate at Edwards Air Force Base.

In recent flights at the High Energy Laser Systems Test Facility (HELSTF) at White Sands Missile Range in New Mexico, Lightcraft have reached a height of 75 feet, with upcoming flights projected for as high as a half-mile or more. The long-term goal of the program is to propel small "picosatellites" into orbit. Such satellites would be about 1 to 3 feet in diameter, with a mass on the order of a kilogram. After being laser-powered to a speed of Mach 5 and about 20 miles altitude in the air-breathing propulsion mode, the vehicle would escape the remaining atmosphere by changing it to a laser-heated rocket propellant. If we build more powerful lasers, we can certainly achieve higher altitudes. Myrabo and Mead hope to get the funding necessary to rebuild and test a high-powered but "moth-balled" laser known as "Driver."

Lightcraft offer environmental benefits as well as cost savings, says Myrabo. Lightcraft emit nothing more than hot air, and their energy comes from above the atmosphere. And they should be cheap—about \$25,000 per launch, compared to \$40 million for the launch of conventional rockets. According to Mead, "Essentially for the price of a new car, you can launch your own satellite and have it up there during the four years you're paying it off."[48]

Lasers are already being used in surgery and on the battlefield for both physical and psychological warfare, and they are still considered a critical component of any future strategic missile defense system. And more important, as the laser-driven energy systems improve, we can predict that our imaginative scientists and policy planners will apply this new technology to providing mass ground transportation on earth (e.g., high-speed rail), powering factories and cities, and heating skyscrapers and mile-high cities.[49]

HAARP, Tesla, and the Coming Magnetic Storm

Perhaps no project more characterizes the species' efforts to exert control over physical forces than the U.S. government's High-Frequency Active Auroral Research Program, better known as HAARP.

According to official communiqués, HAARP was established to acquire knowledge about how weather conditions in space affect communication, navigation, and power grid systems, including a regionwide electrical production network. Another purported goal includes developing technology for detecting incoming cruise missiles and communicating with submarines. The United States Air Force, Phillips Laboratory, and the Office of Naval Research manage the project. Advanced Power Technologies, a subsidiary of E-Systems of Dallas, which is in turn owned by Raytheon, the mega-sized defense contractor, supplies much of the hardware. The Atlantic Richfield Corporation provides a major portion of the on-site energy for the turbines that power the project from its vast Alaskan natural gas reserves.[50]

The main HAARP facility is located in Gakona, Alaska, about 200 miles north-northeast of Anchorage. The site consists of a large radio-frequency-transmitting antenna and a collection of diagnostic instruments. The antenna is currently being configured to transmit 960,000 watts of radio frequency energy into the ionosphere, an upper portion of the Earth's atmosphere that extends spaceward from between 40 and 50 miles above Earth and reaches over 500 miles. Even though the antenna is not yet completed, members of the public and the media attending the Department of Defense's annual "open house" agree that the site is already visually stunning. When it is completed in 2002, this antenna will boast a series of 180 antenna towers that are 70 feet high, in a grid of 15 columns by 12 rows, spaced at 80-foot intervals, creating a complex, interconnected structure of wires and beams.

Project HAARP is the source of both awe and trepidation. These antennae can send out a highly focused and "steerable" electromagnetic (EM) beam that can superheat and actually lift sections of the ionosphere. (The antenna is often referred to as an "ionospheric heater.") The EM waves are targeted to bounce back to Earth from "virtual" mirrors and lenses. They are created by warming specific areas of the ionosphere until they produce a flat or curved shape, capable of strategically redirecting significant amounts of electromagnetic energy.[51]

According to a Joint Services Planning Document issued by the Air Force and Office of Naval Research, HAARP's uses include providing communication to deeply submerged submarines, geophysical probing, controlling the properties of radio waves, and generating mirrors which can be exploited for long-range, over-the-horizon surveillance purposes, including the detection of cruise missiles."[52] Other military documents point out HAARP's potential for altering weather patterns for defensive tactical measures; exploring Earth-penetrating tomography (used to locate weapons facilities underground); and geophysical probing for natural resources like oil, mineral deposits, and gas.

HAARP and its associated technologies could help us control and improve the weather and enhance global communication. According to the claims of one patent, the device would be able to alter "upper-atmosphere wind patterns . . . so that positive environmental effects can be achieved. For example, ozone, nitrogen and other concentrations in the atmosphere could be artificially increased."[53] Theoretically, HAARP could liberate us from the whims of natural disasters and inconvenient climate patterns. It could create rain in drought-ridden areas and decrease it during flooding. HAARP could also redirect tornadoes, hurricanes, and monsoons away from heavily populated areas. Think of the thousands of lives that this new technology could save, the damage it could prevent!

UCLA physics professor Alfred Y. Wong believes that HAARP could be a tool for improving the environment. Wong is the director of the high-powered active-stimulation ionospheric heater, a version of HAARP located in Fairbanks, Alaska. Wong says that HAARP could be used to move chlorine atoms into interplanetary space, thereby minimizing the chlorine atoms' ability to degrade Earth's vital upper atmospheric ozone layer. Wong thinks that we could then control global cooling and warming trends.

Although not originally conceived as such, it is not difficult to envision how ionospheric heaters and other such devices could serve as catalysts to the vitalization process. The ability to change weather and atmospheric conditions and to chart and analyze the composition of subterranean regions would enable us to make distant spheres and planets suitable for human habitation and colonization.[54]

Although the patent for the ionospheric heater is now owned by ARCO, the basic technology underlying HAARP is actually the brainchild of Texas physicist Bernard Eastlund. Those familiar with his work think that his original patents indicate the purposes for which HAARP will eventually be used. In fact, the title of one of the key patents filed by Eastlund could be a brochure description of HAARP itself: "Method and Apparatus for Altering a Region in the Earth's Atmosphere, Ionosphere, and/or Magnetosphere." Other heaters are in operation around the globe, principally in Puerto Rico, Norway, and the Soviet Union. However, the military's "HAARP Fact Sheet" says that HAARP's power levels are 10 times higher than any transmitted on Earth. Eastlund's ionospheric heater is more potent because unlike these devices, its radio frequency radiation is concentrated and focused to a single point in the ionosphere. This difference allows Eastlund's device to throw an unprecedented amount of energy into the ionosphere.[55]

One of the interesting aspects of this patent is that in the patent application itself Eastland actually attributes the fundamental ideas behind the invention to another scientist. He credits the work of Nikola Tesla in the early 1900s for providing the basis for his ionospheric heater.

Nikola Tesla is a shadowy figure in American and world scientific history. Born in Serbia in 1856, Tesla arrived in America in 1884 and by the early part of this century became the most respected inventor in the United States, overshadowing even the renowned Thomas Edison. He was awarded hundreds of patents, including ones for wireless transmission and fluorescent lighting. Tesla made possible the electrification of our modern society—he owned the patents for the alternating current (AC) system of electric power transmission, the breakthrough that enables electric power to be transmitted over long distances. He is also the inventor of the Tesla coil, widely used today in radios, television sets, and other electronic equipment for wireless communication. In 1900, with financing from J. P. Morgan, Tesla began construction of a wireless world broadcasting tower on Long Island. The system Tesla envisioned would provide worldwide communication and furnish facilities for sending messages, pictures, weather warnings, and stock reports. Alas, a financial panic and ensuing labor problems led J. P. Morgan to withdraw support from this project, putting an end to Tesla's dream of a radio communications empire.

Though Tesla never received the Nobel Prize (some suspect professional jealousy as the culprit), Tesla did in fact garner the Edison medal in 1917, the highest honor awarded by the American Institute of Electrical Engineers. The level of adulation and awe that greeted Nikola Tesla in the scientific circles of that time is evident in B.A. Behrend's paean that he publicly read to Tesla at an AIEE meeting in New York City in 1917:

> Nature and Nature's laws lay hid in night;
> God said, "Let Tesla be," and all was light.

In 1900, Tesla made what some consider his most important discovery—terrestrial stationary waves. He proved that the earth could be used as a conductor and would be as responsive as a tuning fork to electrical vibrations of a certain pitch. He lighted 200 lamps without wires from a distance of 25 miles and created man-made lightning, producing flashes measuring 135 feet. He was influenced throughout his life by what he called a "new dynamic theory of gravity." In a speech in 1938, five years before his death, he referred to this theory and announced that he would soon explain it in a more structured format, revealing the "cues of this force and the motions of the heavenly bodies under its influence." His explanation of the dynamic underlying the universe's operation challenged Einstein's concepts about curved space. Some feel that if Tesla had made his theories public at the turn of the century, when he evidently first developed them, his and Einstein's theories might have merged in the scientific literature.

Throughout his life, Tesla would periodically utter some rather strange and boastful pronouncements. He asserted that he could split the earth like an apple, control the weather, affect people's brain patterns; he even claimed that he had developed the theoretic underpinnings of a "death ray" capable of destroying thousands of airplanes 250 miles distant. Although the wild nature of these claims have to this day hurt his credibility among some academics and establishment scientists, many in the military take Tesla's assertions quite seriously. Government and military leaders considered his work valuable property—after his death the FBI, as well as the KGB, were quick to confiscate and classify as top secret any remnants of it that they could get their hands on, including blueprints, drawings, and papers.[56]

Ironically it may be Tesla's more bombastic discoveries that serve as the basis of the HAARP technology. Dr. Nick Begich and Jeanne Manning in their book, *Angels Don't Play this HAARP: Advances in Tesla Technology*, clearly link Tesla's more "creative" ideas to Eastlund's HAARP inventions. Eastlund's patents even reference *The New York Times* articles quoting Tesla's statement that "It is perfectly practicable to transmit electrical energy without wires and produce destructive effects at a distance." Eastlund's patent application echos Tesla's words, claiming that his heater could be used for "military or aircraft destruction, deflection, or confusion" and "weather modification."[57]

HAARP is not without its detractors. Many critics complain that HAARP seems to be driven primarily by the needs of military defense and offense. They point to an internal government document obtained by the magazine *Popular Science*, which clearly states that a major program goal is to "control ionospheric processes in such a way as to greatly improve the performance of military command, control, and communications systems."[58] Early patent documents show that this ionospheric technology could serve as the basis of a "full global shield" to destroy ballistic missiles. The radio signals, acting on the ionosphere, would overheat the missiles' electronic guidance systems as they fly through the powerful radio-energy field. (Dr. Edward Teller, father of the H-bomb and formulator of much of the 1980s Strategic Defensive Initiative [nicknamed "Star Wars"], is rumored to have helped formulate and champion HAARP in the late 1980s.)[59]

To be fair, many counter that using microwave-based technologies for national defense is justified. After all, if the United States and other countries developed such systems, at worst the world would be measurably freer from the threat of nuclear attack.

The bulk of criticism aimed at HAARP centers more around suspicion that no one, including the scientists associated with building the device, can adequately control all effects of these microwave beams on the ionosphere, the environment, or humans, regardless of the benign intentions of its operators. In its 440-page HAARP environmental-impact statement, the Air Force admits that the heater's upwardly directed transmissions could have some effect on objects and people situated under the area of the heated ionosphere. The consequent atmospheric changes might raise the internal body temperature of nearby people, scramble aircraft communications, and ignite road flares in the trunks of cars.[60] (Not unexpectedly, technology buffs and conspiracy theorists have suggested that HAARP-based meddling in the ionosphere might be linked to a number of "unexplained" events, including "Gulf-War syndrome," the crash of TWA Flight 800 in July 1996, and the 1998 disappearance without a trace of 2000 homing pigeons engaged in a long-distance race over the Southern U.S. states.)

Some researchers have even concluded that pulsed radio frequency radiation transmitted over large geographical areas could disrupt the mental processes of people inhabiting such regions. A report released by the International Red Cross raises the possibility that this new technology could disrupt telecommunications and induce negative biological effects. In 1999, a European Parliament Green Paper specifically expressed concern that the "Antarctic HAARP Programme" might damage the environment.[61]

According to HAARP's inventor, Bernard Eastlund, there is no cause for alarm. "What's up there now is not, in my opinion, a big concern—yet. It has to be used judiciously, but it's not the kind of power level that can do the stuff that's in my patents yet. But they're getting up there." HAARP's as yet unconstructed Stage III, comprised of 360 antennas emitting a maximum of 1.7 gigawatts (1,700,000,000 watts) of power, will be able to create "virtual mirrors" in the sky capable of astounding scientific and military feats.

HAARP and other such enterprises demonstrate just how far the species has progressed in its ability to influence and control the dynamics of nature. Both tampering with the ionosphere and overwhelming the forces of magnetism strike fear, awe, and admiration in observers. Although apprehensive about this science's intentional or accidental misuse, we secretly desire to gain control over the weather, search for minerals hidden from human view, and master gravity and magnetism.

Do we really have any alternative but to expand our knowledge of the principles underlying these forces of nature? It would be almost "inhuman" to neglect to gain this knowledge and use it to achieve positive outcomes, for the species and the universe. Progress demands that we develop our potential to manipulate the environment for the common good. Nikola Tesla said as much in a 1940 *New York Times* interview. In the piece, he hinted at the breakthroughs he was making in his own experiments, which many feel set the stage for the technology underlying the HAARP project. One can only take solace from his statement that this new technology "is founded upon a principle that means great things in peace." We must also heed his warning that it can also "be used for great things in war."[62]

The final application of microwave and beam technology will be determined not so much by the technology itself but our values. And if the species is ultimately motivated by the need to achieve vitalization, that is, to bring life, order (as in modifying the weather), and growth to regions and spheres, it will use its new powers wisely and judiciously.

Dominionization's Timeline

Researchers, futurists, sociologists, and policymakers regularly make educated guesses as to the approximate timeline for these new technologies to roll out. Although many innovations have already appeared, others are on the drawing board, waiting to clear regulatory obstacles, while others currently reside exclusively in the minds of their inventors.

George Washington University launched its *Forecast of Emerging Technologies* in the early 1990s to examine the rate and timing of the introduction of information technologies, especially computers. In its later iterations, the researchers enlarged its focus to include all scientific developments, relegating information technologies to a smaller, though critical, part in human development. As they have said in a recent report, "We are not dealing simply with an Information Revolution but with a *Technology Revolution.*"[63] Using diverse methods such as environmental scanning, trend analysis, scenario building, and Delphi surveys of experts in the fields, *The Forecast of Emerging Technologies* constructs timelines and scenarios that predict what technologies will appear and when.

Their latest survey selected 85 emerging technologies for analysis. The *Forecast* predicts that fusion power will be in use by 2026 and sees fuel-cell cars in use by 2016. According to the *Forecast*, we will routinely use biotechnology to produce new strains of plants and animals by 2008; and foods produced through the science of hydroponics should be commonly available by 2015. It sees some form of nanotechnology in use in computers by 2016. I was surprised to see no mention whatsoever of solar-powered satellites in their study.[64]

An underlying assumption of the George Washington University study, as well as others, is that technology develops according to its own internal timeline—technologies will appear on the scene as fast as people can invent and perfect them. If fusion power is commercially feasible by 2016, such studies predict it will be available to the public around that time. In reality, predictions about the technological future must take into account the truism that successful introduction of any technology depends to a great extent on social, cultural, economic, and political factors. The development of fusion depends not so much on overcoming technological obstacles as on the willingness of government and industry to fund this new form of energy. Moreover, if the public considers this technology dangerous, it will pressure the government and the commercial sector to at least delay temporarily the introduction of fusion power.

In addition, when dealing with the future, bear in mind that our species is ultimately unpredictable, prone to magnificent bursts of creativity and innovation that accelerate human progress by quantum leaps. Our species is also most productive when it is driven by a sense of purpose. In the future, as the members of the species become aware of the grander purpose of human existence alluded to in this book, the rate of technological development will accelerate. Driven by a need to control and achieve dominionization over nature, *our* species will increasingly strive to control the power of light itself, the atom, and electromagnetism. That is one reason that I concentrated on research and development in what seem at this moment somewhat arcane subjects, the laser and HAARP. Both these seminal areas went unmentioned in the George Washington University study.

In addition, the exponential growth in the knowledge base is a true wild card in all forecasts. The general population is constantly upgrading its skills, and has more tools at its disposal to acquire knowledge and do some inventing on its own. Nevertheless, most futurists usually look to the established centers of research for the next breakthroughs: government agencies, medical research facilities, and corporate-sponsored laboratories. By focusing on the work of the scientific establishment, most futurists were caught by surprise by the appearance and subsequent rapid spread of the Internet as a form of entertainment and communication. The small group of maverick computer "nerds" who invented the Internet was not even on most futurists' radar screens.

Moreover, it is becoming difficult to predict the speed of the development of technologies and the rapidity of their application. Modern communication tools, including the Internet and wireless transmissions, are reducing the time it takes for information about a new invention or breakthrough to be widely shared. Now, news about innovations such as cloning and nanotechnology spreads almost instantly. Moreover, the impact of these technologies is felt more quickly. For example, cloning, which we will deal with at length in Chapter 2, was successfully performed in a Scottish laboratory in 1997. Within less than two years, the process was being replicated and improved by researchers around the world, and it was applied to different organisms and species.

We can say with certainty that the rate of technological innovation will increase exponentially. And this rapid increase of our knowledge base will help the species master nature. Dominionization is within reach of our species.

As we have seen in this chapter, dominionization is one of the powerful forces spurring our species on to vitalize our planet and the universe in general. However, if humankind is ever to fulfill its destiny, it must learn to master other forces as well.

2

Biogenesis

Creating the Prototype for the Next Human Being

As we enter the third millennium, we are rapidly gaining control not only over our external physical environment but the biological sphere as well. Breakthroughs that encompass a wide range of developments—cancer vaccines, the recent discovery of the immortality gene, genetic engineering, tissue regeneration, cloning, and bionics, to name just a few—will enable us to produce a smarter, stronger, and more durable human being. One new field, *nanomedicine*, which incorporates the principles of the magical new science nanotechnology, will put us in position to shape and redesign the human body as we deem necessary.

I label the sum total of these activities *biogenesis*, an emerging force that will significantly impact the future of the human species. Biogenesis will empower us to control our very physical development and long-term evolutionary advancement and might ultimately lead to the creation of a new human being!

The popular media regularly feature articles that reflect the genuine yearning Americans and others possess to seize control over our biological destiny. A recent *Time* magazine that ushered in 2000 cheerfully chronicled just how far we have come in our centuries-long search for good health and extended life. According to *Time*, science will enable us to live to 200 years old, make our children more intelligent by applying genetic engineering technology, and rid ourselves of cancer, diabetes, stroke, and Alzheimer's disease. Robotics will improve cardiac surgery so much so that soon heart operations will become "failsafe." Advanced computers will usher in an era of foolproof medical diagnosis. The magazine even predicted that people suffering with spinal injuries currently considered irreversible, like the actor Christopher Reeve, might some day walk again.[1]

The term *biogenesis* refers to a number of separate but related activities. It includes the actions that humankind takes to learn and master the fundamental principles of biological growth itself, such as cell growth, tissue generation, and stem cell behavior. However, biogenesis also involves the biological modification and perfection of *other* species as well. Cloning, for instance, is already enabling us to develop stronger, healthier copies or reproductions of sheep, pigs, and mice. These new forms of animals are already a source of spare body parts for humans and serve as a site for humankind to harvest much-needed medicines such as antibiotics that prevent disease in humans.

Biogenesis implies a new relationship between man and nature. From this point forward, humankind, not nature, will determine the shape and direction of human physical development. Let us now explore the breakthroughs that will help us refashion the human being.

Genes and the Human Future

Shakespeare suggested that our fate lies not in our stars, but in "ourselves." More specifically, our future lies in our genes. If we could gain control over our genes we would be able to direct the future development of the human body and begin to rewrite the very laws of evolution. We are rapidly completing the first step in this process, creating a map of the entire genetic makeup of the human being. Once we have this genetic map, we will be in a position to develop methods to manipulate that genetic structure, and thereby modify and completely transform what is the current human being.

Mapping the Genes

The Human Genome Project, a consortium of laboratories and research institutes financed largely by the National Center for Human Genome Research in Bethesda, Maryland, has as its ultimate goal the construction of a blueprint of the genetic composition of the human being. This blueprint will reveal the exact sequence of the four chemical letters making up each one of the 80,000-plus genes of the human species. Moreover, this blueprint will show the exact location of each of those genes on the 23 pairs of chromosomes.

This grand image of the human genome will help researchers determine the function of every gene in the human body. More importantly, this basic information will help us discover the genetic causes of many diseases. Geneticist Eric Lander of the Whitehead Institute for Biomedical Research in Cambridge, Massachusetts, claims that with this map, "You can take this huge genetic database of, say, people with heart disease and determine which form genes associated with the heart have taken. That will tell you what went wrong, and then you can make therapies based on that."[2] Many diseases are caused by imperfections in the genes. It is hoped that once the Human Genome Project has identified the specific genes related to a particular disease, medical teams can then proceed to use this knowledge to cure or prevent gene-related diseases. For instance, some believe that individuals get cancer because certain genes are *not* activated properly. The Human Genome Project will identify the problem gene or gene cluster. We can then develop techniques to correct the flaws in these genes so that we can eventually prevent the development of cancer in individuals unlucky enough to have a genetic predisposition to it.

Originally researchers were to complete the project by the year 2010. However, breakthroughs in the very methods used to study the human gene have opened the possibility that the project will be finished several years earlier than expected. In late 1998, Francis S. Collins, director of the National Human Genome Research Institute announced that the sequencing of all of the 80,000-plus human genes might be completed by 2003.[3] Yet even as he reported the good news, the private sector weighed in with a loftier timetable. In early 2000, Dr. J. Craig Venter, head of the firm Celera Genomics, a private partnership, claimed that his company would identify the entire human genome sequence as early as summer 2000 and no later than 2001, well ahead of the Federal effort (and for hundreds of millions of dollars less).[4] Venter, long known as a maverick in the world of molecular biology, uses a cutting edge gene-mapping computer known as a "gene sequencer" to decode the human genome at lightning speed.[5]

Finding the genes that cause disease is the first step in actually curing the disease itself. Discovering these linkages might also help medical teams diagnose diseases with more accuracy. If we can discover early a person's susceptibility to a certain disease, we can then proceed to suggest therapies, lifestyle changes, or dietary programs that can prevent the person coming down with the disorder.

For example, in mid-1997, a research team under the direction of Dr. Davis Parker, Jr., of the University of Virginia Medical Center in Charlottesville, working in collaboration with Mitokor Inc. of San Diego, discovered the genetic link to some 60 to 70 percent of Alzheimer's disease cases. Researchers concluded that defects in genes that regulate energy metabolism in the cell could lead to a simple diagnostic test for the disease.[6] Genetic research may also enable us to prevent blindness in many cases. Recently scientists at the University of Michigan Medical School and in New Zealand, France, Germany, and elsewhere identified three genes that when flawed lead to the disease retinitis pigmentosa, which causes severe vision loss. The disease, which afflicts 100,000 to 200,000 Americans, strikes people at a fairly early age. The suspect genes are instrumental in processing sight-giving vitamin A. If we can use genetic screening to identify this early in a person's life, we might be able to save his sight if he is administered high doses of vitamin A early in his life. At the same time, researchers announced that they had located a gene involved in macular degeneration, a leading cause of blindness in older people originally thought to be a natural result of aging. With our new knowledge we can begin to devise genetic, dietary, and other therapies to cure the disease or preempt its onset.

Incidentally, the above case illustrates how one force, species coalescence, helps the species in its quest to marshal the forces of biogenesis. The synergistic efforts of an international team comprised of researchers from the National Institutes of Health, the National Center for Human Genome Research, and their colleagues in laboratories in Italy led to the discovery of the site of a gene believed to be the cause of many varieties of Parkinson's disease. This team successfully mapped the general site of the suspect gene, chromosome 4, which causes the neurodegenerative disease that afflicts about 1 percent of the U.S. population over 50 years old, or about 1 million to 1.5 million Americans.[7]

As the third millennium begins, we are continuing to discover new links between genes and diseases. Researchers at Vanderbilt and the University of Washington in Seattle have found the genetic routes of 90 percent of all breast cancers.[8] Lawrence Berkeley National Laboratory's Human Genome Center in Berkeley, California, found the genetic roots of Down's syndrome, a disease that afflicts 250,000 Americans.[9] Researchers at the University of Colorado and the Veterans Affairs Medical Center in Denver have found some of the genetic defects that lead to a brain abnormality causing a schizophrenic's inability to filter out irrelevant audio and visual stimuli.[10] Icelandic researchers found the genetic basis of severe hand tremors. This disorder, common in Scandinavian countries, causes the victim's hands and head to shake and interferes with his ability to write and eat.[11] In late 1998, researchers at the University of Texas Southwestern Medical Center in Dallas reported that they had located a gene that helps to regulate the growth of cells. They claimed that when this gene mutates, is damaged, or malfunctions, it might help promote lung and colon cancer. This discovery adds to a growing list of gene mutations linked to cancer.[12] As we entered 2000, Britain's Glaxo Wellcome Plc announced that it had succeeded in identifying the human genes that could be linked to migraine, diabetes, and psoriasis.[13]

Fixing the Genes: The Promising World of Gene Therapy

The genetic approach to disarming and neutralizing diseases is moving ahead at a breakneck speed. The Human Genome Project is the first stage in this battle—identifying genes and their functions. The second stage, which some think will achieve total success by 2020 or earlier, involves delivering these genes into the human being to correct both the maladies themselves and an individual's vulnerability to developing a specific disease. Simply defined, *gene therapy* is the placement of beneficial genes into the cells of patients. According to Inder Varma, a professor at the Salk Institute in La Jolla, California, by introducing the gene and consequently the protein it produces, "you either eliminate the defect, ameliorate the defect, slow down the progression of the disease or in some way interfere with the disease."[14]

The first actual use of gene therapy began in September 1990, as a treatment for a child suffering from a deficiency of the enzyme ADA (adenosine deaminase). People with this deficiency are prone to persistent infections and risk early cancer, and in many cases die in the first months of life. The root of the disease is the body's inability to produce a key chemical because of a defect in their gene coding; people who inherit the defective gene from both parents stand a good chance of acquiring this disease. Children can only survive if they live inside a totally germfree environment (leading to them being labeled in the popular media as "bubble children"). In the 1980s, researchers began infusing children with ADA deficiency with ADA gene-corrected cells every one or two months. Suddenly, these patients could lead normal, happy lives outside of their restricted bubble environments. This type of gene therapy, whether for ADA deficiency or other diseases, requires regular transfusions—the altered cells will not last forever. A permanent cure would require bone marrow stem cells, which we discuss elsewhere. However, at most their plight is no worse than that of diabetics, who receive daily injections of insulin to manage their disease.[15]

In a recent case, an innovative application of gene therapy saved a Texas heart patient's life. In spite of the fact he had undergone a bypass operation for his chronic case of angina, this condition had returned with a vengeance. Living on nitroglycerine tablets, he was basically on a deathwatch, too incapacitated to undergo another heart-bypass operation. Then he heard that St. Elizabeth Medical Center in Boston was conducting clinical trials of a new genetic therapy for heart patients. He volunteered for the trial, a decision that saved his life.

At the medical center Dr. Jeffrey Isner and his team injected a solution containing billions of copies of a gene that activated blood-vessel growth directly into the patient's heart. The therapy worked in this Texan's case— his heart is regrowing its blood vessels, he is off the nitroglycerine, and he has regained his active life. All 16 heart patients in Isner's trial have showed improvement; six are entirely free of pain.

Presently, over 300 such clinical trials, with over 3000 participants, are in progress. Why, then, is gene therapy not progressing more quickly and not already a standard medical procedure? Salk's Inder Varma sums up the problem. "There are only three problems in gene therapy: delivery, delivery and delivery. It isn't going to be a problem to make gene therapy work—if we have an appropriate set of tools to deliver the genes." While the strategy, substituting a normal gene for a faulty one, is seemingly simple, scientists found that adding genes to cells could impart new, and sometimes unwanted, functions to those cells. You do not want to induce other disorders, such as heart disease or cancer, while eliminating the target disease, say, cystic fibrosis. The cure, in this case, would be as bad or worse than the disease.

Therefore, we must discover ways to get the gene into the cell without upsetting the normal functioning of the body. Researchers have tried to use altered viruses to deliver genes into the body. However, this is not a problem-free method—the body often resists and totally rejects the virus itself, neutralizing the impact of the improved implanted gene the virus is carrying.[16] In 1999, Ariad, a Cambridge, Massachusetts, biotechnology company working with the University of Pennsylvania, devised a new type of gene therapy that circumvented such problems. Ariad's method involves first injecting the patient with the modified gene, which would remain dormant until the patient takes a pill to turn it on. As a result, when new genes are delivered into the body, scientists can monitor their side effects, including wholesale rejection by the body of the gene. This new method promises to provide us a way to quickly turn off the newly implanted genes in case a patient developed an unanticipated side effect. Ultimately, human patients might be able to get a gene injection every couple of years, then merely take a pill occasionally in order to activate the new gene. In this way the precise dose of the needed protein could be raised or lowered as needed.[17]

The future role of gene therapy in the biogenesis process is promising. Gene therapy is already being used in trials conducted by GTI-Novartis to attack and destroy brain tumors. In the near future we will use gene therapy to cure cystic fibrosis, leukemia, herpes, arthritis, Huntington's disease, and sickle-cell anemia. The possibilities seem endless. Soon we will be introducing "suicide genes" into breast or ovarian cancer cells, or into cells infected by the AIDS disease virus, in order to make them vulnerable to attack by the bodies immune system or drugs. As 2000 began, the Cleveland Clinic Foundation announced that it would start clinical trials to use gene therapy to combat prostate cancer in men whose conditions resisted other forms of treatment. Annually over 40,000 American men die of this disease.[18]

Some 40 biotech companies, including Novartis Pharmaceuticals, Merck, Johnson and Johnson, and Amgen, are racing to bring gene therapy products to a worldwide market, which may reach $3.6 billion in sales by the year 2005. Dr. W. French Anderson, director of gene therapy at the University of Southern California Medical School, predicts that "twenty years from now gene therapy will have revolutionized the practice of medicine."[19]

Designer People: Reengineering the Future Human

Since 1990, the technique underlying gene therapy has been the placement of a healthy gene into the cells of a single organ of a patient who suffers from a genetic disease. However, a new day is dawning! Soon we will alter a fertilized egg so that the genes in every one of a person's cells will carry a gene that scientists or technicians have placed there. In essence, genetic engineering will enable us to not only change and improve what we are now, but determine what the next stage of human development will be.

We will acquire this new capability by mastering a biological technology labeled "germline engineering." *Germline* refers to the eggs and sperm. The purported use for germline therapy would be to shield a future child from his family's predisposition to some genetically determined disease, such as cancer, cystic fibrosis, or sickle-cell anemia. According to James Watson, president of the Cold Spring Harbor Laboratory, a discoverer of the structure of DNA and once head of the Human Genome project, we owe it to humanity to "take away the threat of Alzheimer's or breast cancer." In the long run, germline therapy might actually be a more efficient way to attack the genetic roots of a disease than gene therapy. Currently, to combat a disease such as cystic fibrosis, we implant a new gene into millions of lung cells in a cystic fibrosis patient. By contrast, germline therapy would involve manipulating the fertilized egg of a potential disease victim before he or she is born, thereby enabling the gene to reach every cell of the person born to the mother possessing that egg.

A recent magazine article aptly described how germline therapy will better the human condition. The writer asks us to picture the plight of a family whose males are genetically predisposed to develop prostate cancer. The man carrying the genes for prostate cancer desires to rid his progeny of the curse of this disease forever. He and his wife visit a clinic specializing in germline therapy. Doctors there fertilize a few of the wife's eggs with her husband's sperm, and then inject an artificial chromosome into the fertilized egg. This chromosome contains made-to-order genes carrying instructions that instruct the prostate cells to self-destruct. The doctors then place the embryo into the woman's uterus. If her baby is a boy, when he gets old he might very well develop prostate cancer. But the cell suicide gene implanted in his mother's egg before his birth will make his prostate cancer cells self-destruct. Moreover, not only will he not die of prostate cancer; the new cancer-killing gene has copied itself into every cell of his body, which includes his sperm. Therefore, his sons will not die of cancer either, nor will his sons' sons.[20]

The long-term effects on the human species are obvious. As we gradually eliminate the many causes of death, we will naturally be able to live much longer than we do today.

Some think that we are only 10 to 20 years away from realizing the dream of eradicating genetic illnesses. The public is quickly coming to realize that genetic science may hold the key to the elimination of breast and ovarian cancer, cystic fibrosis, ALS, and other horrors that have plagued humanity since time immemorial. They will soon pressure governments, medical associations, and various "ethics commissions" to lift regulations hampering the development and application of such technologies and therapies. In a poll taken at a recent UCLA symposium on genetics and health, two-thirds of the audience supported using germline engineering to eradicate disease.

Our manipulation of the human genome may eventually transcend merely eliminating disease. We may eventually utilize genetics to determine the physical composition of our progeny, including their height, body type, hair color, and skin tone. We will probably be able to prearrange their intelligence level, perhaps their personality type, and most certainly their sex. In 1998, researchers at the Genetics & IVF Institute in Fairfax, Virginia, demonstrated that we already can use genetic science to determine the sex of our progeny. They successfully helped parents select the sex of their children, using a technique known as a *mechanical sperm sorter*. All the couples in the study wanted to have girls, since at least one member of each couple carried a genetic tendency toward a deadly disease that almost always affects boys; 93 percent of the wives participating in this project gave birth to girls.[21]

How prepared is society to adapt to these changes? If one judges by the media's reaction, any attempt to improve the human species through genetic manipulation smacks of eugenics. The 1997 movie *Gattica* portrayed a society in which one's genetic makeup is one's destiny. *Gattica*'s society permitted only a limited number of people to be programmed for excellent health and superior intelligence. Others were genetically programmed to have the capacity to fill only the menial and secondary roles in society—human street cleaners, housekeepers, and security guards. Genetic science was the tool by which an authoritarian elite rules a joyless planet. (To reinforce their antigenetic message, the producers incorporated an even more insidious symbolism—all the members of the movie's genetically determined elite are Caucasian. In other words, the producers subtly correlated genetic engineering with racism.)

The general population, according to *Time* magazine, possesses a much more favorable attitude toward genetics than do the media. In a recent survey, clearly 60 percent of the respondents would choose to use genetic screening to rule out a fatal disease of their baby. One-third would choose to genetically endow their baby with greater intelligence. While only 10 percent of the respondents claim they would try to genetically influence the child's height, weight, or sex, one can speculate that if the technology were available now, most potential parents might be tempted to ensure that their offspring be endowed with high intelligence, good looks, and resistance to all major diseases. Modern society is fairly addicted to anything that promises physical improvement, including cosmetic surgery and vitamin therapy. I doubt that people would deny their offspring the same luxuries they would without hesitation bestow upon themselves.

There are numerous apprehensions about the germline engineering process. Will the experiments get out of control? Will the new genes introduced into the sperm and egg function in ways that we did not intend? Will the process work too well, leading to production of a generation of "superpersons" who make the older generation of "normals" superfluous, archaic, and hopelessly inferior? Government action certainly reflects this tentative feeling toward the new technology. While government maintains a laissez-faire attitude toward most forms of DNA manipulation, it closely regulates any experiment that involves the *inserting* of genetic material into human germ cells. Policy makers recognize that such insertions would be the first step in redirecting human evolution into unknown areas, however supposedly beneficial.

Dr. Henry I. Miller, senior research fellow at Stanford University's Hoover Institution, wrote in a *Wall Street Journal* article that society will ultimately accept the use of gene therapy for purposes beyond the elimination of disease. Society will turn to genetic engineering for "enhancement," the increase in human physical or mental capacities. According to Miller, applying gene therapy to physical enhancement differs little from the current use of drugs to treat obesity, age spots, and baldness. We are engaged in clinical trials of appetite suppressants, memory- and performance-enhancement drugs, and human growth hormone for short children. Of course, genetic engineering may have a more powerful impact than such treatments on our mental and physical capacities.[22]

Recently, James Watson emphatically stated that we should "never postpone experiments that have clearly defined future benefits for fear of dangers that can't be quantified." He feels that we can react rationally only to real risks, not hypothetical ones. To do otherwise would be to condemn us to a static, stagnant social and physical environment. And truth be told, ultimately our species, when given the choice, will choose perfection and improvement. On the individual level, humans' natural tendency to favor physical enhancement will cause them to accept germ therapy. Society as a whole may eventually employ germline engineering as the normal procedure in the procreation process.[23]

In the future, the spacefaring human species will constantly have to physically adapt ourselves to new environments of new planets. How better to do this than through manipulation of the human genetic structure, our own and future generations'? To successfully meet the complex challenges that a spacefaring society will confront, we may have to reengineer the brain and the nervous system. Future generations just might be required to be a great deal more intelligent (or more "brain efficient") than even the brightest among us today.

At the Brink of Immortality

For all of humanity's existence, achieving life everlasting has seemed like nothing more than a beautiful dream. Now, new research may expand our biological capabilities to the point that we soon may be able to equip the human body with a set of instructions to simply stop aging.

We are on the verge of finding the holy grail of all biological research—the gene or set of genes linked to aging itself. Once we locate that "aging" gene, all we must do is find a way to control it. At that point, we will not only cure disease; we will be empowered to *reverse the very aging process itself.*

In 1997, Geron Corporation, a small biotechnology company located in Menlo Park, California, announced that it had located the gene implicated in the human aging process. To appreciate Geron's breakthrough, let's quickly look at how the human cell matures. Cells divide as the human body ages. Each time the cell divides, it sheds tiny bits of DNA labeled *telomeres.* It would be to our benefit to retain these telomeres—they protect our chromosomes, much like caps on shoelaces protect the lace. These telomeres are key to prolonging our lives and our health. Unfortunately, after perhaps a hundred cell divisions, these telomeres shorten and hence lose their ability to protect the cell. As a result, the chromosome becomes damaged.[24]

If we could find a way to keep our telomeres intact, we could theoretically live forever. The enzyme that lengthens the telomeres, protecting and rebuilding them, is known as telomerase. When this enzyme is expressed in a cell, that cell is for all practical purposes "immortal"; many biologists have dubbed telomerase the "immortalizing enzyme" because of its power to bestow "life everlasting" on cells.

Not unexpectedly, for several years researchers had been looking for the gene responsible for the production of this powerful enzyme telomerase. Scientists at Geron Corporation and the University of Colorado at Boulder discovered (and quickly proceeded to clone) this gene, labeled hTRT. With this gene in hand, researchers now possessed the means to produce the protein that provides telomerase's blueprint.

In January 1998, Geron announced that it had activated the enzyme telomerase by inserting hTRT into a cell. The experiments succeeded as planned: The Geron scientists were able to extend the treated cell's telomeres and lengthen the cell's life span *in culture* by at least 20 divisions past the so-called natural limits of cell division. Later that year, Geron announced that in a separate experiment it had reconstituted the telomeres of embryonic stem cells, the cells known for their ability to turn into any type of cell at maturation. If Geron could make such stem cells immortal, the species would be in position to possibly rejuvenate parts of any organ with a simple injection.

Ronald Eastman, Geron's president and CEO dedicated his company's resources and energy to discovering a way to introduce telomerase activity into mortal cells in order to extend their replicative life span.[25] Some fear that if we inject the enzyme into the body, the body's blood system would treat the enzyme as a foreign body and quickly destroy telomerase, long before it could do its intended job. Science writer Ben Bova envisions the development of an artificial, man-made form of telomerase that could withstand the body's defense systems while still maintaining the essential characteristics of natural telomerase. This synthetic telomerase would be injected into the bloodstream, resist the body's rejection system, and begin to rebuild telomeres on the cells.[26] An alternative suggestion would be to engineer genes that produce telomerase, as Geron Corporation has done, and proceed to put them into a virus or other similar carrier.

While the process seems promising, there are some nagging questions regarding both the experiments and the antiaging concept itself. Although each cell in the human body possesses the gene for producing telomerase, our cells do not attempt to express the gene. Why, we may ask, do our cells not make telomerase and let the human being live forever, or at least enjoy some degree of ultralongevity? It would seem to be in the cells' best self-interest to not inhibit the continued activity of this life-giving enzyme.

The answer to this question, Bova points out, reveals one of nature's ultimate paradoxes. The body contains telomerase *inhibitors,* which prevent cells from proliferating uncontrollably. These inhibitors act as part of the body's natural defense against cancer. In other words, the body recognizes that some forms of extensive cell regeneration will end up killing the human organism. Our cells do not express their telomerase gene simply because cells that constantly rebuild their telomeres could possibly become cancerous. This process protects us from developing cancer; unfortunately, it also prevents us from becoming immortal.

We will have to find a way to use telomerase without triggering the production of deadly cancers. Geron and other companies involved in the search for immortality face a major problem during experimentation, especially with humans. If the cells don't produce enough telomerase to regrow the telomeres, the experiment is a bust. The cells will stop replicating and die. However, if too great a quantity of telomerase finds its way into the cells, the patient would grow tumors that will quickly get out of control.

Early indications are that we will be able to "outwit nature" without triggering a medical disaster. In late 1998, the scientific world was abuzz with rumors that some findings favorable to Geron's claims regarding telomerase would be forthcoming. Then, on January 1, 1999, the scientific journal *Nature Genetics* published two papers that concluded that the enzyme could be used to extend the lives of normal cells without causing the runaway-type cell growth typical of cancer cells. In one study, Geron scientists showed that human skin cells and retinal pigment cells exposed to telomerase and injected into mice over a year before had been continually dividing, and *must be considered immortal.* Importantly, according to Geron, "these same cells retain normal growth control and do not form tumors" in the bodies of the mice. The second study, conducted at the University of Texas Southwestern Medical Center, also demonstrated that normal cells exposed to telomerase in laboratory dishes acquired immortality without becoming cancerous.[27]

One caveat ought to be mentioned here about this "miracle" being created in the laboratories. Even if science discovers how to extend cell lives indefinitely, such a technology will not *reverse* the aging process for anyone undertaking such genetic treatment. That job will be the domain of other methods, like skin, tissue, and organ regeneration, or nanotechnological productions of new body parts. The Geron-style cell manipulation will not enable a 40-year-old man to regain his 20-year-old body; however, it will prevent him from aging past 40, because his cells will not age or die. It is safe to predict that once this technology has been perfected and its application is widely available, most people at age 20 or 25 will routinely begin whatever treatments necessary to stymie the cell-aging process. After all, who would not want to stay young forever?

This ability to overcome the aging process in the human being carries tremendous ramifications for the species. We can assume that longer lives will afford all of us the opportunity to become not only older but wiser— we will mature but not age. We will acquire a greater sense of the "big picture," the interconnection between the past and the future. And immortality will give each of us a greater stake in the future. After all, if I know I will still be around 300 years from now, I will do what I must to ensure that the future is a healthy and prosperous one. In a world where immortality is the norm, *the* future is *my* future.

Biogenesis and Nanotechnology

In order to perfect the human species and master the biological system as a whole, we must apply the full measure of our knowledge to develop advanced technologies that will help us in our cause. One such cutting edge technology is nanotechnology.

Recall that the goal of nanotechnology is to control matter at its most basic level, that of atoms, molecules, and electrons. Although this science is still in its infancy, many observers are already speculating that when perfected this technology will help us cure human diseases and rebuild damaged human bodies.

Ralph C. Merkle, a director of the aforementioned Foresight Institute and also a key researcher in the nanotechnology area with the Xerox Corporation, sees nanotechnology playing a major role in the defeat of disease and ill health. According to Merkle, disease is caused largely by damage at the cellular and molecular level. Nanotechnology, which Merkle calls "the manufacturing technology of the twenty-first century," will empower us to build fleets of computer-controlled molecular tools much smaller than a human cell and as accurate and precise as a drug molecule. These atom-sized machines will allow us at long last to intervene in a controlled yet highly effective way at the cellular and molecular level.

Nanotechnological machines have two very distinctive advantages when dealing with the repair of the human cell—the nanomachines will be extremely small and exceedingly smart. In the future we will construct nanomachines that can identify and kill cancer cells without impairing or impacting neighboring healthy cells. Imagine a small device equipped with a minicomputer and a supply of some "poison" that could be selectively released to kill a cell the device has identified as cancerous. The nanomachine would circulate freely throughout the body, locate cancer cells, and follow its programming to release specific poisons when it came in contact with specific cancers. Engineers could monitor the nanomachines' movement and behavior and reprogram them via acoustic signals whenever they deemed it necessary to do so.[28]

Science writer Ben Bova sees nanotechnology as a gateway to human immortality. Bova speculates that disorders as diverse as cancer and degenerative nerve diseases, obesity and atherosclerosis, could be treated by swallowing a glass of orange juice containing hundreds of millions of nanomachines preprogrammed to treat an assortment of bodily ailments. In his vision, these nanomachines—atomic- and molecular-sized robots—spread throughout the body, rebuilding worn tissue, organs, and muscles, strengthening failing bones, keeping the arteries clear of obstruction, and engaging in other life-saving and life-preserving activities. Imagine these nanomachines acting as an ultratiny surgical team destroying tumors wherever they find them.[29]

Foresight Institute founder K. Eric Drexler contends that once we master the principles of nanotechnology, we should be able to so thoroughly manage human biology that we eradicate disease and aging. Drexler sees disease as a molecular phenomenon, a situation caused by various crucial molecules being out of place. For instance, sickle-cell anemia is caused by a single specific amino acid erroneously located in the structure of hemoglobin. An individual plagued by this error of nature cannot process oxygen normally. Drexler, like others, envisions an army or fleet of tiny programmed robots streaming through that individual's bloodstream, repairing such cellular displacements, in essence curing the person of the disease. In Drexler's nanoworld, all diseases, even aging itself, will be cured through the miracle of nanotechnology.

Nanotechnology will be able to cure aging simply by retarding the slow accumulation of errors in the chromosomes residing at the core of cells. Biologist Christopher Lampton claims that we could design a nanosubmarine that could "proofread" a person's chromosomes for possible errors and repair such errors before they accumulate to a fatal level. He sees this "nanosub" equipped with six "proofreading devices" that would be attached to it on long molecular cables. Each device would enter a cell and read the cell's chromosomes. Its central nanocomputer would compare the chromosomes and look for differences in the molecular sequences. When the nanocomputer finds a cell that contains a chromosomal sequence different than any of its neighbors, the nanosub corrects the sequences. Amazingly, Lampton declares that a human body subjected to such nanotechnological probing and repairing would slowly revert to its optimum state, defined as a body of an individual in his or her mid-twenties. Lampton, like others, sees this treatment initiated by an individual simply swallowing a glass of liquid filled with billions of these nanomachines, which then do their repair work.[30]

To fully control the human evolutionary process, we must significantly enhance our current understanding of the human biological system.[31] Merkle thinks nanotechnology will help us achieve this—he envisions autonomous molecular machines exploring and analyzing living systems, including the human body, in greater detail than ever before. These machines could enter the body, take tissue samples, and bring them out of the body for analysis. These "snapshots" would enable us to acquire a map of the human body system more detailed than any we can currently imagine. Nanotechnology should expand dramatically our comprehension of the complex and often perplexing processes that take place in the human body.[32]

I can imagine programming these nanomachines to carry molecular-size "digital video cameras" that would transmit back to us a genuine inside view of the human body from a vantage point within a molecule or cell. Such a perspective will enable us to arrive at a "picture-perfect" understanding of the structure of the amino acids and other entities in the body. We could even incorporate *virtual reality* technology into this process—a researcher could take a virtual journey into a DNA molecule and see and feel first hand how the molecule operates and interacts with the rest of the biological system.

And remember—nanotechnology is first and foremost a manufacturing technology. This will certainly help us achieve vitalization. Nanotechnology will enable our species to create wholly new organs for greater adaptation to any environment. Imagine our future spacefaring descendents growing different types of breathing apparatus to survive comfortably in a novel planetary environment or modifying their skin composition to overcome the effects of more intense suns in distant solar systems.

Cloning and the Perfection of Humanity

Biogenesis represents the full spectrum of human efforts to improve biological systems. Humankind guides its physical development and evolution. In addition, we are gaining control over the biological forces regulating all living matter, such as cell growth and tissue generation. Biogenesis also includes the application of this knowledge and techniques to the development and perfection of the nonhuman living organisms.

Cloning, the exact replication of cells, body parts, organs, and whole organisms, is a science that meets all the above criteria. Before 1997, cloning occurred only in the movies or in science fiction pulp magazines. In the 1996 movie *Multiplicity,* a man, overchallenged by his busy work and family schedule, decides to clone himself (with the help of the usual mad scientist). The protagonist, aptly played by Michael Keaton, hopes his clone can serve as a fill-in on some of his job- and family-related activities. The experiment goes awry, and by movie's end Keaton finds himself in the company of several Keaton-clones, most who have their own ideas about how his life should be run.

As it turns out, although *Multiplicity* treated the idea of cloning as a source of humor, the joke was on all of us. Before the movie could complete its inevitable odyssey from screen premiere to video, cable, and commercial television release, cloning made its own transition from fantasy to reality. In February 1997, a breakthrough occurred that literally stunned the scientific community and the public at large. A research team at Edinburgh's Roslin Institute, working under the leadership of embryologist Ian Wilmut, successfully cloned a sheep, creating the sheep's exact genetic duplicate. Scientists had previously cloned animals such as frogs and monkeys from embryonic cells. The Scottish team, however, cloned an animal from a fully developed adult cell.

It took the participation of three different ewes to produce Dolly, the cloned sheep. In the Roslin cloning process, the embryo that became Dolly received all her genes from one individual, a 6-year-old adult ewe. The researchers removed the nucleus containing a set of genes from a mature udder cell of the 6-year-old ewe and placed this nucleus into the egg of a second ewe whose nucleus had been removed. The researchers manipulated both the egg and transplanted nucleus so they would develop into an embryo. They placed this newly created embryo into the uterus of a third ewe, which carried Dolly and later gave birth to her. Dolly, whose genes are an exact replication of those of the 6-year-old ewe that contributed the udder cell, is that 6-year-old's exact clone. She received no genes from a male father nor from the other two ewes involved in the cloning process.

Dolly was a perfectly healthy sheep, so healthy, in fact, that in early 1998 Dolly herself became a mother. In 1997, the Roslin Institute put Dolly through a breeding program, mating her with a male sheep. The public's skepticism about cloning abated as it became clear that Dolly, though a clone, was like any other living organism, physically and genetically sound, fertile, and unaffected by her own rather unusual conception.[33]

The reception to the experiments that created Dolly the sheep were decidedly negative. In July 1997, the World Health Assembly labeled the cloning of humans "ethically unacceptable." President Clinton placed a moratorium on Federal funding for any experiments that involved human cloning until a commission could sort out the ethics involved. The National Bioethics Advisory Commission encouraged private researchers to voluntarily comply with the U.S. government's moratorium on human cloning experiments. Hiroshi Nakajima, director of the World Health Organization, took a softer approach, noting that animal cloning can improve diagnosis and treatment of human diseases.[34] Israel flatly banned human cloning for five years. Israel's Parliament will reconsider the ban in 2003.[35]

Clearly, the underlying fear regarding cloning is the possibility that some scientist or research team would attempt to replicate a human being. After all, the Roslin experiments were being successfully imitated using a variety of species, including mice and cows. How long before some laboratory or research team would try to reproduce an exact replica of a human?

The waters of public opinion were tested in early 1998 when Richard Seed, a Chicago scientist, announced that he wanted to establish a clinic to pioneer a cloning process that could produce offspring for infertile couples. He announced, some felt prematurely, that he would within 18 months have mastered the technology to produce a clone-based pregnancy. Seed's methodology would follow the Roslin model closely, leading to the birth of an exact genetic identical twin of the mother. He claimed that he had already located four couples willing to participate in his experiments, and he boldly stated that he would consider opening the clinic in another country if U.S. law or courts prevented him from practicing the cloning arts in the United States.

Seed's announcement was greeted with concern and in many cases alarm. Lord Robert Winston, a London-based fertility expert involved in the first "test-tube" baby in 1978, said that "there is no way you could clone a human being safely at this point." Princeton University President Harold Shapiro, chairman of the National Bioethics Advisory Commission, was quoted as saying that the ethics issues surrounding the cloning of humans are unresolved and that perhaps in the future such a procedure will be feasible, even desirable, but not now. Seed replied that it would be unethical not to explore this new science's possibilities. In a press statement, Seed claimed that humans were created in God's image. God, he said, intended humans to become like Him and achieve a godlike state, and to be godlike we must strive to be immortal. "Man should have an indefinite life and have indefinite knowledge," Seed said, "and we're going to do it, and this (the cloning of humans) is one step."[36]

It didn't take long for Seed's speculations to get dangerously close to reality. In December 1998, a South Korean research team stunned the public by announcing that it had cultivated a human embryo using an unfertilized somatic cell, the type that makes up most of the body. Citing Lee Bo-yeon, a researcher with Kyonghee University Hospital's infertility clinic, the report went on to say that researchers at the clinic had actually cloned a human cell. The researchers used cells donated by a woman in her thirties to cultivate the embryo, which then divided into four cells. Lee said that at that point they aborted the experiment.

However, had Lee and his colleagues actually taken the next step in the process and implanted the embryo into another woman, the embryo would have grown into a child with the same genetic characteristics as the mother-donor! American cloning experts said it was the first time they knew of that human DNA had been transferred from a body cell into a human egg, with the egg then developing into embryonic cells. Body cells, as opposed to eggs or sperm, contain the full complement of a person's DNA. It became clear that it was only a matter of time before someone would implant cloned human cells and place them in a human birth surrogate. At that point, we would have our first baby clone.[37]

The reaction to the announcement by the South Korean researchers was quick, definitive, and stark. The South Korean government immediately banned government funding for research into human cloning. But this was a "soft" ban—the government acknowledged that it could not prevent "maverick doctors" from forging ahead with human cloning experiments. The government would not fund such research, but would take no legal action against the scientists. The legislature promised to pass legislation that would "not punish anybody (for cloning humans) but will have a strong warning effect on the few scientists who are interested in cloning research," according to Rep. Rhee Shanq-hi.[38]

In downtown Seoul, 20 civic activists demonstrated, brandishing signs that transmitted a strong message: Ban human cloning research now! Cloning evidently offended many Korean's deeply rooted Confucian beliefs. (One person carried a sign with a row of identical mug shots of himself. The caption underneath asked "Which one is the real Me?") American Life League President Judie Brown issued a statement that the South Korean claim to have "created" a cloned human embryo—a human being—"ought to send shock waves down the spines of all Americans who have thus far remained blind to the consequences of man's insistence that he is God." Ms. Brown claimed that "reproductive technology has been on a steady slide toward Frankenstein's nightmare." She called for a congressional ban on any scientific experiments which she claimed "jeopardize the integrity of the human being, whose life begins at fertilization." Jeremy Rifkin, an antitechnology activist, stated that "We are determined to mount a global effort in opposition to human cloning." He also called for immediate legislation to ban such experiments.[39]

As we enter the first years of the twenty-first century, as far as we know no one has yet accomplished the complete cloning of a human being. However, one might rightly suspect that the strong public reaction to the South

Korean scientist's success might have driven the research on human cloning underground. Hence, one cannot be sure that human clones do not at this time already exist. Let us remember that the South Korean team aborted their research out of a fear of success, not failure.

Nevertheless, the practical benefits of cloning will very soon outweigh the negative reactions. If we can precisely control genes, we can engineer farm animals that can produce more milk, meat, or wool. Cloning would allow us to develop large herds of animals such as goats and cows who have already been genetically altered to produce in their milk certain rare and valuable medicines and pharmaceuticals. National Institutes of Health director Dr. Harold Varmus believes that cloning can teach us how to turn genes on and off. Perhaps we could induce human genes to produce new tissue for the repair of old or diseased tissue—new limbs, skin, even bone marrow.[40]

Breakthroughs reported in April 2000 suggested that through cloning technology the human species might achieve a form of superlongevity. A company called Advanced Cell Technology (ACT) announced that it had cloned calves whose cells were biologically younger than cells of normally conceived newborn calves.[41] ACT's breakthrough has led some to speculate that soon we will be able to clone a generation of humans who have a much longer life span than we currently enjoy.[42]

We already see the possibility of using cloning for human organ replacement. A biologist at ProBio, Dr. Tony Perry, wants to develop a bank of healthy animals with usable parts that could be harvested for human organ replacement on demand. Of course, this goes to the epicenter of the philosophical debate that will become part of the "battle for the future." Some people are already saying that we have no right to use animals in this way, regardless of how such technologies may benefit the human species. Perry responds: "To them I say, I believe strongly in the ascendancy of humans. We should do everything we can to save people."[43]

The cloning process is becoming ever more safe and efficient. Japanese researchers claim to have devised a highly efficient cloning procedure, which enables them to create eight calves from the cells of one adult cow. And the Japanese team was successful in 8 of their 10 attempts. By contrast, the Scottish researchers who cloned Dolly took 277 tries to create one lamb. We must develop a highly efficient, zero-defects cloning technology if we hope to successfully produce on a large scale genetically identical livestock with desirable traits (such as the ability to produce high-quality milk).

We even see cloning used to fulfill the other meaning of biogenesis, the perfecting of not only the human species but also the preserving and "shepherding" of other species. China was originally one of the major critics of cloning in general. (The Chinese Academy of Sciences originally wanted human cloning banned entirely.[44]) Now, Chinese scientists may turn to cloning to save the giant panda from extinction. Currently, only 1000 pandas survive in the wild, and their numbers are decreasing. The plight of the panda has now induced the Chinese Academy of Sciences to publicly endorse cloning as a method to increase the population of the regal panda.[45] Some scientists think we might use cloning techniques to bring back long-extinct species. Japanese scientists were reportedly attempting to clone a mammoth, a prehistoric elephantlike mammal, out of eons-old genetic remnants of this extinct creature. This inspired scientists in Thailand to initiate a project to clone a white elephant that belonged to a nineteenth century Siamese king, Rama III. The remains of the elephant, believed to possess the finest characteristics of a male adult elephant, have been preserved. The Thai biologists will use the genetic material from Rama III's prize elephant to clone a new one, a process they expect to take 10 years to complete.[46]

The public will accept cloning when science shows that this new technology can save human lives and cure our illnesses. This may occur sooner than we think! The University of Texas Medical School cardiologist Dr. Ward Cassells envisions using the genes of a child suffering from an incurable disease, like leukemia, to clone a child who could provide his older twin lifesaving bone marrow. More controversial would be his suggestion that we use cloning technology to produce genetically identical embryos of children as a source of "spare parts" for failing organs in the living child.[47]

When we begin to clone humans, we will be rewriting the rules of evolution. We will now begin to select, in much the way nature has, the primary traits and genetic tendencies that we consider most important for our survival and ultimate progress. Christian Crews, writing in *The Futurist*, represented this view beautifully:

> We are about to begin a century of biotechnology that could give us in the next few decades what natural evolution may never provide. It would take hundreds of millennia for evolution to adequately adapt our bodies for the rigors of extended space travel. Using genetic manipulation, we could do it in a generation.[48]

We are certainly entering a brand new era of human achievement!

The Many Faces of Biogenesis

The biogenesis process includes developments even more dazzling than achievements in genetic engineering, nanotechnology, and antiaging technologies. The human species is remodeling and reengineering itself, by regenerating skin cells, manipulating the very basic stem cells, and creating artificial versions of human organs.

Bionics: Making the Artificial Real

Historically, humanity has replicated parts of the body that through disease have become dysfunctional or through accident have been irreparably damaged. Common examples are eyeglasses, false teeth, hairpieces, and artificial limbs. These parts have become increasingly sophisticated. Artificial joints, hips, ankles, and knees, are quite common, and the heart pacemaker is one of the modern forms of electronic wizardry that helps keep people alive by regulating their heartbeat.

However, we are now entering an era in which we are combining the computer with other advanced technologies to create a new category of medical innovation labeled bionics. The newest developments in this field are quite dramatic.

For instance, patients with chronic heart failure usually have to depend on heart transplants to continue their lives. The success of heart transplants depends on several factors: the availability of the actual organ, the success of the operation itself, and the extent to which the body accepts the new heart. Now, there may be an alternative to organ transplants. At five U.S. hospitals, surgical teams are preparing for a groundbreaking operation to replace failing hearts with battery-operated mechanical pumps made of plastic and titanium. Although surgeons are only just testing these artificial hearts on calves, they hope to test the new totally implantable artificial hearts in human patients around 2000 or shortly thereafter. These artificial hearts are similar in size to the natural human heart. Developed at a number of sites, including Penn State University, 3M, and the Texas Heart Institute, they run on various machinelike mechanisms such as pumps, chambers, and pistons. They are truly bionic, regulated by computers, internal censors, and microprocessors. The hope is that soon 60,000 heart patients a year will be able to use these implanted devices.[49]

While germline therapy and nanotechnology may be the major contributors to human health and well-being, artificial appendages and organs will play an important role in the biogenesis process. We will turn to such bionic devices to ameliorate the damage to organs that occurs through accidents and other unintended events. Moreover, as such devices radically improve, many of us may choose to replace our natural organs with mechanical ones. After all, who wouldn't want "super vision" of 20/10 power or a heart that would last forever?

The Coming Age of Tissue Reengineering

Biogenesis is a force that will serve as a catalyst to a continual rebirth and re-creation of the human species. Certainly, the ability to regenerate our very own bodies through *tissue engineering* measures up to this lofty aim. This scientific craft seeks to discover techniques that will coax the body to regenerate damaged tissue and missing parts.

With the right technology tissue can be regenerated anywhere on the body—scientists at University of Massachusetts and MIT dramatically demonstrated that principle when in a burst of creativity they ran an experiment in which they grew a human ear on the back of a mouse. At the forefront of this new science are companies such as Advanced Tissue Sciences, Inc., and Organogenesis, Inc. In the late 1990s, Advanced Tissue Sciences developed several products that resemble human skin. One of their products, Dermagraft, will when approved be used to repair the skin of burn victims.[50] Another product, Apilgraf, received an unexpected spurt of publicity when it was used to treat a deadly skin disorder of an 8-week-old baby girl. The young lady, Tori Cameron, was born with a condition called *epidermolysis bullosa*, in which the top layers of skin do not adhere to each other. The form of the disease plaguing Tori is rare and can be fatal in the first two months of life. The pressure caused by the very process of being born caused Tori to lose about 80 percent of her skin to blistering in the first days of her life. So fragile is her skin that the slightest bump or pressure can cause it to blister and fall away.

Scientists and dermatologists performed a series of applications of the specially made skin. Apilgraf, produced by Organogenesis, is created from skin cells nurtured in culture dishes, which are then developed into a dermal layer. (The skin cells originally come from donor baby boys' foreskins donated by parents following their sons' circumcisions.) Doctors are slowly replacing much of Tori's dermal layers with the bioengineered skin. As of early 1999, they had replaced about 40 percent of her original skin with Apilgraf. According to Dr. William Eaglestein, chairman of dermatology at the University of Miami, "the hope is that the cells (in the replacement skin) will teach the baby's skin to act normally." Tissue therapy might save Tori Cameron's life![51]

Tissue and cell engineering or regeneration will enable us to replace faltering organs. One of the pioneers in the field of tissue regeneration, Dr. Joseph Vacanti of the Children's Hospital in Boston, is working on the basic science involved in taking a few healthy cells from a sick person and using such cells to grow the patient a new nose or ear. His brother, Dr. Charles Vacanti, Director of the Center for Tissue Engineering at the University of Massachusetts Medical Center in Worcester, is attempting to tissue-engineer jawbone, ears, and even a trachea. One of his most miraculous accomplishments involved regenerating a patient's thumb, which had been smashed in a machine accident. He extracted cells from the patient's arms and nurtured the cells in a gel for six weeks. Then he took a piece of sterilized coral, which is similar in composition to bone, and sculpted it into the shape of a thumb bone. Vacanti's team injected the matured cells into the coral, and then placed the coral at the end of the patient's injured thumb. They hope that the coral will dissolve and in its place living bone will thrive once more. This process has worked in other cases using ears, chests, and other body parts.

Scientists at other institutions are developing tissue-engineered corneas, new pancreases for diabetics, and breast tissue for mastectomy patients. A research team directed by Dr. Michael Sefton of the University of Toronto intends to grow an entire heart within the next decade. Sefton's vision is to "be able to pop out a damaged heart and replace it as easily as you would replace the carburetor in your car."[52]

On the conceptual level, tissue engineering science seems challenging. However, once the species has perfected this technology, laypersons possessing the proper materials, training, and a user-friendly manual, should be able to perform organ-replacement operations without doctors or engineers present. The implications of this new technology for the vitalization process is obvious—space colonists millions of miles from Earth would be able to perform some body replacement operations on severely injured members of their teams. The more independent space travelers are, the more successful their expeditions will be.

The Miracle of Stem Cells

Possibly no development will more dramatically advance the biogenesis process than the progress researchers are making to exploit the power of stem cells.

Every cell in our body contains instructions to make a complete human. Most of these instructions are inactivated, and for a very good reason—if these instructions were active, brain cells, for example, might start producing stomach acid. A nose could potentially turn into an ear.

However, fetal cells, during the earliest stages of pregnancy, actually have the potential to turn into any and all body parts. At that point, the fetus' cells are *stem cells*, which have not yet begun to specialize. For some time scientists have recognized that such stem cells have the potential to create healthy tissue in adults. Most diseases involve the death of healthy cells—brain cells in Alzheimer's, cardiac cells in heart disease. It stands to reason that if we could isolate and control stem cells, we could provide patients with healthy replacement tissue. We have already successfully used such undifferentiated fetal stem cells to help Parkinson's disease victims. Even though adult brain neurons normally don't regenerate, stem cells taken from fetal brains could grow and replace adults' damaged brain cells, apparently without harm to the recipients. A team lead by Dr. Ronald McKay of the National Institute of Neurological Disorders and Stroke transplanted fetal brain cells into some 200 patients with Parkinson's disease. The results were miraculous. According to McKay, patients who were unable to move could now walk around their houses, even swing a golf club without interference.

Before long, we won't have to acquire such cells from fetuses. In late 1998, a research team led by Dr. James A. Thomson at the University of Wisconsin, Madison, and a Johns Hopkins University research team led by Dr. John Gearhart independently cultured human stem cells. Now science can anticipate culturing an endless supply of human stem cells that have the ability to develop into a wide variety of human tissues.

While application of this breakthrough may be more than a decade away, Gearhart and Thomson projected that these stem cells could be used to grow nerve cells to repair spinal injuries and restore function to paralyzed limbs. Stem cells could be routinely used to make brain cells that would secrete dopamine for the treatment and control of Parkinson's disease, as McKay's team had done. We might be able to grow heart muscle cells to replace useless scar tissue after a heart attack. We could make islet cells that produce insulin and create a lifelong treatment for diabetes. With stem cells, we could even manufacture blood cells that are genetically altered to resist specific diseases, such as HIV.[53]

Daniel Perry, executive director of the Alliance of Aging Research, spoke of a development of a whole new field, *gero-technology.* "Gero-tech is when medical science uses new insights about the aging process to develop novel processes and therapies that ultimately could help cure, postpone or prevent age-related diseases." According to Perry, once we can maintain self-renewing yet long-lived colonies of human cells, we will apply cell transplantation techniques to the treatment of a myriad of disease, including breast cancer and heart disease.[54]

Evolution on Our Terms

In this chapter, we have only scratched the surface of the many breakthroughs that will lead to the fulfillment of the biogenesis process. Even as this book goes to press, members of our species are feverishly working to invent vaccines for cancer and other "incurable" diseases. Others are working on therapies to regenerate damaged spines—in April 2000 Boston Life Sciences, Inc., successfully stimulated regrowth of severed spinal cord motor nerve fibers using a nerve growth factor it had developed called Inosin.[55] In addition, we are making much progress in the field of *cryonics,* a relatively new science that seeks to maintain diseased or injured human beings in a "frozen" state so that they can be unfrozen at such time as a cure is found for whatever ails them.

These advances strongly suggest that humankind is quickly taking charge of its own evolution. However, before I show how we are commandeering the very forces of evolution, let me say a few words about this theory itself.

As a concept, *evolution* has been with us for over almost two centuries. Although we largely associate this theory with Charles Darwin, scientists and naturalists writing before the publication of Darwin's *The Origin of Species* had been speculating on the methods by which humanity reached its current physical state. What made Darwin's theory of evolution so convincing was the solid data he presented to buttress his arguments. In 1831, Darwin embarked on a five-year global voyage on the HMS *Beagle.* That voyage afforded the young naturalist the opportunity to amass a wealth of biological and geological specimens from all over the world that he would eventually weave into his famous theory. The basic thrust of Darwin's theory of human development still prevails in the orthodox biological sciences.

From his evidence, Darwin concluded that all life forms developed from, or grew out of, one or a few very primitive species, and that all species *evolved* from simpler forms to more complex ones. The basic process underlying this progression of the species is *natural selection*. Species evolve when through chance mutation new traits appear. Of all these new trait variations, the ones more favorable to the species' continued survival and enhancement are retained by the species, and the useless variations are discarded. For instance, suppose during the evolutionary process one member of the species mutated and exhibited a new trait, such as stronger limbs or better eyesight. According to Darwin, the natural selection process would mandate that over time the species would retain these newly acquired superior traits.

How do these traits arise in the first place? Why does one member of the species suddenly exhibit these superior traits, such as stronger bones or sharper hearing? Modern evolutionary theory maintains that the changes that occur in the human genetic structure underlie most transformations in the human physique. These changes, these mutations, are caused by changes in the organism's DNA, a rearrangement of the DNA molecule. These mutations in genetic makeup produce physical variations in the organism. If the new traits caused by the genetic changes help the mutated member of the species gain certain advantages over the nonmutated members of the species, those new traits will eventually predominate throughout the species.

Modern evolutionary theory deduces that genetic changes that make mutant members of the species faster, bigger, and healthier and that improve their eyesight, posture, and manual dexterity will also enable those organisms to survive longer and breed faster than others. Because the "new and improved" organisms are passing on these new traits faster than the nonmutated, and for a longer period of time (they live longer), their genetic mutations gradually take over the gene pool. Eventually, those organisms become the majority variant after one or more generations.

As you can probably infer from our examination of humankind's foray into the biogenesis process, we have placed ourselves in a unique position to upend these very fundamental mechanisms underlying natural selection. Evolution is basically the process by which favorable genetic variations are preserved and unfavorable ones destroyed. Suddenly, as we enter the third millennium, the human species now possesses the ability, first, to decide what are "favorable" or desirable characteristics, and, second, to manipulate our genes to induce the manifestation of these traits in individuals.

It is not so much that the process of evolution and natural selection is no longer valid. Rather, in the twenty-first century, we are taking the principles of evolution and commandeering them for our own purposes.

The processes we have discussed—manipulation of the stem cells, nano-technologically based transformations of the human biological system, and genetic engineering—all mimic the process of evolution that Darwin considered exclusively "natural" and in many ways accidental. Indeed, we are on the verge of taking the randomness out of our own species' evolution, as well as that of other organisms. We will use germline engineering to permanently change the course of human development by preselecting the genetic structure of future generations. When the mass of the population opts for progeny with high intelligence, great manual dexterity, and resistance to a host of diseases, we will be using the process of "planned selection" to permanently change the human genome. Those possessing the superior genes will be healthier and live longer than those without such genetic enhancements. Therefore, over time their unnaturally induced traits will "naturally" predominate throughout the human gene pool. Eventually the human genome, the blueprint that defines us as human, will permanently change.

The human species, in its efforts to elevate itself and nature, is expressing a propensity to perfection that has been present in humanity since its earliest history. The biogenesis process represents a multitude of activities, all performed willfully and with real goals in mind: taking control of our own evolution, influencing the evolution of other species, eventually transposing that evolution to other spheres and worlds, and fostering other planets' organic evolution. Our species' ingenuity and intelligence even has us knocking on the door of immortality.

Moreover, we will not only adapt to our environment through such processes as biogenesis, but also change the environment on this planet and throughout the universe to make it more user friendly to the human species. Through dominionization we are literally turning orthodox evolutionary theory on its head: Adaptation becomes a reciprocal process in which humans change to adapt to new environments, but physical environments on Earth and elsewhere will be changed to facilitate the presence of human beings.

Throughout the nineteenth century, philosophers and naturalists promoted the idea that humanity intentionally strives for perfection. Nietzsche considered *Will* (his capitalization) a kind of animate force directing man and nature. In fact, Nietzsche directly accused Darwin of overstressing the environment as a shaper of man and other beings. In contrast to Darwin's position, Nietzsche believed that humankind has an innate ability to shape and create new forms. Schopenhauer posits the existence of a "will," an unconscious, purposeful, irrational force that manifests itself in the natural world. And George Bernard Shaw referred to a *life force* that could create and organize new tissue to achieve its goals. It would surprise many of us today that Shaw actually considered *Darwinism* not so much a science as a form of mythology!

Evolutionists, as well as others arguing within the so-called *post-modernist* framework, are loath to consider *will* as a force in evolution. Randomness, accident, chance mutation, and certainly lack of any overriding purpose underscore the modern evolutionary paradigm. The concept of "will," that is, the knowledgeable, conscious striving on the part of life to create and re-create itself in ever-higher forms, is repudiated by most modern Darwinians. Brian L. Silver, in his encyclopedic and brilliant *Ascent of Science*, sums up this position succinctly:

> It is essential to realize that evolution is an automatic process; it involves no "will" on the part of the organism involved, and the occurrence and type of a given variation, unless human beings deliberately intervene, is governed by chance. There is no instinctive drive in a living organism to create "higher" forms of life. It is the combination of chance variation and the environment that gives evolution a direction.[56]

The *expansionary* vision of the universe, which I will explain at greater length in Chapter 6, summarily rejects this purported "triumph of mean-inglessness." We can see the influence of human will in the very actions described in this chapter and throughout the book. We are pushing, creating, changing, and transforming everything around us. The modern evolutionists counter that the actions of twentieth and twenty-first century humans to intervene in the evolution of life and the planet are a special case, anomalous behavior unlike human actions over the species' first million years of exis-tence. The *expansionary theory* would argue that throughout recorded history (from the Sumerian era to the present), humankind has been expressing its will to succeed, progress, and change. We have ceaselessly been adapting to our environment, making what we consider the right choices for our survival and ultimate growth.

Of course, the central question driving this book is *why?* What motivation, what force, conscious or otherwise, ultimately drives humankind to intervene in the natural forces of biology? One could argue that we pursue medical breakthroughs and expend our resources on medical, scientific, and health-related research exclusively to improve the comfort and health of the human species. Certainly, we are utilizing modern medicine, genetics, and pharmacology to eliminate illness and make our lives a more enjoyable experience.

However, on closer inspection we notice that our activities in the area of biogenesis enable us to achieve much more than good health. Through biogenesis we are establishing physical properties and capabilities that are uniquely equipping us to achieve the goals of vitalization as outlined throughout this book. Several facts strongly suggest that humankind unconsciously pursues biogenesis to enable our species to ultimately vitalize the universe.

First and foremost, the vitalization process will require a version of the human being who becomes, over time, "enhanced." To vitalize the universe our species must operate at its optimum efficiency—more creative and intelligent and less prone to physical degeneration. With the help of many of the advances outlined in this chapter, we will produce a human being more durable, adroit, and resistant to disease than the current version. We will continue to engage in this perfection of the human species as an ongoing work in progress for many eons. In fact, centuries from now as we migrate across various quadrants of the universe, we will be continually improving ourselves into a more perfect human, a version of *Homo sapiens* even more skilled than we are in the art and science of enhancing our world.

Second, vitalization will require that we travel to other spheres. Many of these new environments will be initially uninhabitable by humans. Certainly, our skills in dominionization will help us transform the environments of others spheres to become "human-friendly." However, in some cases we will have to genetically alter ourselves in order to survive on new planets. The mere act of traveling in space presents a myriad of challenges to the human body. The problem of prolonged weightlessness in space is only beginning to be understood. Living on spacecraft during lengthy interplanetary and eventually interstellar voyages might mandate that we genetically alter our explorers' bodies. Some have suggested using cryonics as a way to "deep-freeze" space travelers whose voyage extends beyond the current human life span. They would be "unfrozen" upon their arrival at their distant location (another star system, perhaps), as young and spry as the day they left Earth, ready to begin their task of colonizing, mining, and exploring. Moreover, if we are to live on other planets, our skills in the area of cloning will come in handy. Cloning will enable us to replicate body types resistant to diseases and adaptable to a variety of space environments. Therefore, it is important that we learn how to modify the human body, the physical shell.

Third, vitalization will require a human being that can survive longer than the current model can. Some missions we embark on might last for decades, possibly even centuries. Such projects would require individuals who can see the whole project through to its termination. In other words, we must discover a way to achieve ultralongevity or near-immortality in humans. Such long-lived members of a mission will develop a sense of connectedness to their respective projects that we "temporary employees" have never known.

Lois Wingerson, in her book *Unnatural Selection*, examines the impact that genetic engineering and related technologies will have on human development. She states that "if any force is driving changes in the environment these days, it is ourselves." The qualitative change between past eras and the present is that now "we not only define progress, we either create it or confound it by our actions and our interactions with our surroundings." She concurs with our view that we now have a capacity we never had before—to direct our own evolution by "manipulating human genes" and other scientific advances.[57]

Those who fear that by tampering with natural evolution we are incurring formidable risks to the species and the planet ought to keep such errors in perspective. Evolutionists and modern biologists routinely overglorify the wonders of the natural evolutionary process. In fact, an inspection of the human body shows that such products of natural evolution are far from perfect. Every day we discover genes that because they act imperfectly and are prone to disrepair lead to disease and other bodily malfunctions. Nature's greatest mistake, in fact, might be that it has left us mortal.

As the twenty-first century begins, we are modifying the body at the genetic and molecular levels, transforming ourselves into physical beings undreamed of by nature. We are becoming the beings who can accomplish goals scripted by the human imagination who will bestow our knowledge, consciousness, and humanness on distant spheres and serve as the catalyst to the future shape of the universe.

3

Cybergenesis

Of Men and the Smart Machine

Biogenesis empowers us to defy our evolutionary fate and create our own biological destiny. Through dominionization we challenge and reshape the very forces of nature. These two forces will enable us to modify and influence the development of the human being and the ordered evolvement of the material universe.

A third force is emerging that like the first two will permit us to further human evolution and meet the greater challenges confronting our species. This force, cybergenesis, is the enjoining of humankind and its smart machines—computers, microchips, cybernetics, and a variety of inorganic mechanisms—to improve the cognitive and physical abilities of the human species.

Like the other forces, *cybergenesis* embodies several related meanings. In the first place, humans will extend our intelligence and consciousness outward to the machine. We are creating extrafast, ultra-smart machines such as the supercomputer and the quantum computer, and we will endow with human-like intelligence systems such as information networks and the Internet. The computer will serve as a surrogate memory, enabling us to store amounts of information that the human brain could never hope to retain. It will help us reason more sharply, calculate at speeds and levels impossible at the human-brain level, construct complex simulations of the man-made and natural worlds, and help us in our decision making. To the extent it makes us smarter, and hence more adaptable, the smart machine will become an integral part of our evolution.

Cybergenesis refers to our efforts to not only extend "humanness" to the machine but to integrate elements of the smart machine into the human. We are already using computer-based devices to help our bodies physically function better than they would "naturally"—such devices will improve our normal capacity to hear, see, feel, and touch. In addition, we will someday also use cybernetic technology to enhance the body's cognitive systems, such as the brain. The enormous breakthroughs promised by the exciting research performed at British Telecom, for instance, portend the development of technologies by which human mental powers will be enhanced through chip implantation. With the help of *nanotechnology* and other advanced techniques, we will bring all those wondrous cybernetic powers now residing in external computer systems into the human body by planting inside our physical shell molecular and atomic nanosized computers!

Certainly, if we hope to challenge the laws of physics, overcome entropy, extend human existence beyond the solar system, and re-create a living Earth throughout the cosmos, we must develop a much greater cerebral capacity than we currently possess.

Adding to the Measure of Man

A recent edition of *Time Digital, Time* magazine's occasionally published "Guide to Personal Technology," devoted an entire issue to "Techno Sapiens," an entity borne of what *Time* described as "the ultimate marriage of technology and humans." The issue's cover was certainly intriguing—it pictured a human arm protruding from the side of the page. What differentiated this arm from the normal human appendage was its appearance—a collection of tubes, wires, and batteries encased in a plastic transparent tube representing the forearm, at the end of which stood a plasticene-like hand with a wire attached. The inscription next to the arm stated "whether they're aiding the disabled or extending the reach of scientists, computers are becoming part of us all."[1]

Inside the magazine we see the full picture, and we are shocked. We meet one Ken Whitten, a former utility linesman who lost both his arms in an electrical accident. That strange looking futuristic cover photo is an actual picture of one of his arms, a chip-controlled prosthetic limb. This device, controlled by a built-in computer, has restored to him an "arm" strong enough to perform most normal tasks—Whitten can move the arm and use it to lift objects. With this "cyber-arm" Whitten also feels objects as he touches them.[2]

Whitten's case represents an early example of the emerging trend toward inserting the computerized smart chips into the human being to replace missing or damaged body parts, an intriguing combination of electrical engineering and medical technology.

A key aspect of the *cybergenesis* process is the eventual integration of the computer into the human organism itself. Once the stuff of science fiction, the implantation of computer chips into lower primates for purposes of tracking and monitoring them is now possible. Soon, the amalgamation of computer and human will transcend such mundane applications. Already we have implanted devices into the hearing and visual centers in the brain that allow the deaf and blind to get a modicum of sensory data into those areas. And we are developing prosthetics that can respond to muscle triggers.

That the Blind May See

For centuries man has used prosthetics, specifically artificial limbs, to replace damaged or missing legs and arms. Modern surgical techniques have enabled doctors to routinely place artificial hips into people's bodies to help them retain their ability to walk. Breakthroughs in tooth implantation have enabled us to replace deteriorating teeth without having to turn to uncomfortable dentures. The artificial heart, of course, is already recognized as one of the marvels of the modern age.

Now computer-based technology will help us further enhance the functioning of the human body. For years those with hearing problems have employed computerized hearing devices to improve their ability to hear the sights and sounds most of us take for granted. Recently, cochlear implants have helped people with severely impaired hearing (the near-deaf) to regain the capability to experience the wonderful sounds of our world.

Now, we may be on the verge of using chip implants to overcome blindness, one of the greatest handicaps plaguing humanity. For several years we have been developing a host of new techniques, such as lens implants and corneal surgery, to ameliorate certain sight-related problems. However, the most common cause of blindness, a deteriorating retina, has eluded similar technological fixes. Up to now, once a person's retina, the light-sensing portion of the eye, started to deteriorate, we had little hope to restore that person's vision. Six million Americans suffer from vision impairment, including blindness, because of retinal disease.

Now, thanks to a computer chip, scientists may be able to replace the retina. Johns Hopkins University researchers and University of North Carolina at Chapel Hill electrical engineering professor Wentai Liu are working on a technology that holds out hope to those with severely impaired retinas. They have devised an implant, a specially designed computer chip, that would be surgically placed at the back of the eye, near the optic nerve. The patient wears a specially designed eyeglass frame on which a tiny camera is mounted. The camera scans the environment, takes in the images, translates them to electrical impulses that it sends to this computer chip/artificial retina. The chip then sends these impulses to the brain as pictures the wearer can "see."[3]

Early tests suggest that we may be on the verge of a major breakthrough in the war on blindness. In studies, patients using this artificial retina have been able to detect motion and shapes of objects. One patient using such an implant even identified the first letter of his name written on a card held in front of him.

Researchers in Japan, coming at this problem from a slightly different angle, are developing a hybrid electronic-biological artificial retina. In their model, actual retinal cells are cultured on the back of a light-sensitive semiconductor chip; the chip in turn stimulates the retinal cells' electrical pulses, mimicking the process by which a normal eye actually sees. To many people, these new implements may sound very much like technological gizmos featured in *Star Trek: The Next Generation*. In fact, these are exactly the kind of technologies that American and Japanese researchers are attempting to replicate.[4]

Another method has been developed by Vincent Chow, co-founder of the company Optobionics. Using computer chip-manufacturing techniques, he and other scientists at Optobionics are etching thousands of microscopic solar cells onto a tiny silicon chip in an effort to have these artificial sensors take the place of the natural ones in the damaged retinas of blind people. Like the other cyberchips, this solar version is implanted in the back of the eye. When light strikes those tiny solar cells, the cells turn the light into electrical signals that travel through the optic nerve to the brain and are interpreted by the wearer as an image.

Patients with *retinitis pigmentosa* and macular degeneration will be prime candidates for chip implants. Patients suffering from *retinitis pigmentosa* at first lose peripheral and night-vision, which slowly leads to tunnel vision. Macular degeneration causes a loss of sharpness in the central field of vision and worsens with aging.

One of the major obstacles to the success of "cybervision," as some have labeled this burgeoning field, might not be so much technological as perceptual. The brain of someone who has not experienced sight for many years is actually programmed—wired—differently than the brain of someone who can see. Even after we implant the new retina into the patient, we might actually have to reprogram the blind recipient's brain so he or she can recognize the images that the technology is transmitting. We must stimulate the mind's eye in order to get the brain to see images it is physically receiving.

Some people in the field think that the perception issue will not be a permanent obstacle to the success of the new cybervision technology. "With this process, the brain, when it receives a signal, it should not be able to tell the difference whether it came from a naturally healthy retina or from our implant," Dr. Chow exclaims.[5]

Embedding Chips to Right Nature's Wrongs

In the near future, we will be implanting chips in humans to cure all sorts of ills. In France, a French physician at the Fourier University in Grenoble has found a way to use chip technology to help Parkinson's disease victims who are totally immobile gain mobility in an instant. He is doing this with a brand new science, *neural implant therapy,* in which computer chips controlled by radio or other signals either interfere with or augment the activity of brain circuitry to improve patients' physical or motor abilities.

Parkinson's disease is caused by a curious string of physical factors. All of us have in our bodies a "neurotransmitter" called dopamine. If we do not possess sufficient amounts of this neurotransmitter, two specific tiny regions of our brain become overactive. Such overactivity in these regions negatively impacts how the body functions. Parkinson's patients show such effects: slowness of movement, stiffness of the entire body, even an inability to walk correctly or at all. Eventually, the patient becomes totally paralyzed, and in many cases dies.

Dr. A. L. Benebid at Fourier University found that he could reverse Parkinson's symptoms by inhibiting the overactivity in those two regions of the brain. Paradoxically, Benebid achieves this by actually overstimulating these two brain sectors with an electrode he has permanently implanted in the patients' brains; this electrode is wired to a small electronic control unit placed in the patient's chest. Benebid uses radio signals to program the unit, even turn it on and off.

When he turns the radio waves on, the patients can walk around the room, functioning as normally as could anyone not afflicted with this disease. As soon as the doctor turns the electrodes off, though, the patients' symptoms of Parkinson's disease immediately return. The new neural implant therapy is approved in Europe and awaits FDA approval in the United States.[6]

Other diseases and disorders are being treated with neural implants in what is being called "deep brain stimulation" therapies. Doctors have achieved some success in using such implants to reduce tremors associated with cerebral palsy, multiple sclerosis, and other tremor-causing diseases. They are doing this by implanting electrodes in a section of the brain called the *ventral lateral thalamus.* Clearly, science is beginning to treat the brain like the circuitry system that it is.

We are now using implants to help the deaf as well as the blind. Increasingly, cochlear implants together with electronic speech processors that perform frequency analysis of sound waves are functioning as well as the inner ear in many cases. About 10 percent of the formerly deaf persons receiving this neural replacement device are now able to hear and understand others' voices sufficiently to engage in normal telephone conversations.

Many observers believe that in the coming years all of us, healthy and impaired, will utilize implants to enhance our vision and hearing. We can just imagine how implants could be employed to alleviate the pain endured by patients suffering from cancer and other debilitating diseases. Once we locate the regions of the brain that control pain, we could short-circuit pain sensations with the flick of a button.

Some are even predicting that in the twenty-first century we will be able to communicate with and control machines such as computers by transmitting impulses from these neural implants to these smart machines. Such implants might even permit us to download and upload information into a computer or onto the World Wide Web.

We can gain insight into how this man-computer interface might work from current examples of the use of chip implant technology to eliminate disabilities. At Emory University Hospital in Atlanta, scientists recently implanted a tiny electronic device in the brain of a 53-year-old paralyzed man that enables him to communicate directly with a computer. The electronic device, called a neurotrophic electrode, is coated with certain chemicals that encourage surrounding nerve tissue to form connections with the brain. In a sense, this electrode, once implanted, gradually becomes part of the patient's brain. The electrode picks up the electrical signals running through this connection and wirelessly transmits them to an FM receiver outside the scalp. No wires actually go into the patient's skin.

The patient learns how to use his "brain power" to control the length and pattern of impulses emitted by the electrode. Using these impulses, the patient can actually move a cursor across a PC. The patient, totally paralyzed, can use the cursor to select letters and spell his name, and with sufficient training, compose notes and e-mail them. A world of communication will be opened to a person who had assumed he would spend the rest of his life trapped inside his body, never again to experience the joy of expressing emotions or ideas to the rest of the world. According to Dr. Phillip R. Kennedy, who invented the implant while at Georgia Tech, "The hope is we can enable many paralyzed persons to communicate."[7]

Researchers are developing other methods to establish connections between humans and machines. Over the last several decades various methods of biofeedback have been used to help people learn how to concentrate, relax, relieve migraines, and perform other activities. Currently scientists are performing experiments in which subjects are using their brain patterns, unaided by implants, chips, or other devices, to operate a computer. A Florham Park, New Jersey, quadriplegic is one of five people participating in a worldwide experiment with Cyberlink, a device that magnifies brain activity to the point where the brain can operate a computer. The patient, Bruce Davis, has been paralyzed since he had a car accident when he was 38. At age 52, he will turn to anything to better his condition.

He wears a band around his head which controls three electrodes that magnify the electricity emitted from his head by 500,000 times. That electricity travels via modem into the computer—the modem transforms the electricity into a signal the computer can understand. The user slowly learns what thoughts or mental energy can make the machine perform various screen activities. One type of "thought" can make the computer open an icon on the screen, and another can make the cursor click on the icon that can make the computer log on to the Internet. Since the computer can be connected to any other system, over time quadriplegics could learn how to use this brain-computer interface to adjust their electrically operated beds, turn lights on and off, and perform other activities. Alas, there are only 40 Cyberlink machines in existence on planet Earth.[8]

Although we increasingly will employ neural implants to solve numerous health and medical problems, they are not without some bit of controversy. The concept of implanting any type of intelligent device in the body, be it an artificial retina or a "neural electrode," is bound to be met with suspicion if not outright hostility by many. Peter Cochrane, head of research at British Telecom Labs, links the public's apprehension over brain implants to its deep mistrust of science in general. He blames the media for such anxiety. TV, radio, and the cinema, he says, sensationalize such topics as brain implants, and demonize members of the scientific profession. "One of the greatest disservices being done to us," Cochrane laments, "are these Hollywood movies and Saturday morning cartoons, which present a depiction of us as mad scientists and engineers, always in white coats, cackling madly and trying to take over the world."[9]

Some members of the public feel that the use of chip implants in humans increases the chances that the government might use such technology to practice "mind control" on the population. They contend that any group possessing such technology is armed with potent tools for mass manipulation. Cochrane and other scientists liken such opposition to standard anti-technology sentiments. As Cochrane says, "I'm here just kind of experimenting, playing with a mix of technologies that I believe will benefit humankind."[10]

To overcome the public's suspicion about these technologies' potential for mischief, government agencies, universities, and research institutes involved in funding and implementing such chip implant technologies should adopt a policy of full disclosure. These organizations should share information about the nature and purpose of such technologies with any and all takers, and regularly provide the public reports on the status of such technologies. The more we expose and demystify scientific developments, the less the public will fear and mistrust them.

So far, agencies have been less than forthcoming about chip implant technology. Such secrecy makes it difficult for any interested parties, including writers and journalists, to get a true handle on developments in the field and explain the benefits of these technologies to the layperson. Recently, when reporters for the *Electronic Engineering Times* attempted to acquire information from the National Institutes of Health on the status of chip implant technology research, they were refused interviews with key personnel involved with this research. The NIH forbade Terry Hambrecht, longtime head of the Neural Prosthesis program at the NIH, from speaking to the press on NIH's work with neural implants. Moreover, the agency would not provide the press any background information on the work.[11]

The purpose of all scientific activity is not only to advance scientific knowledge and development. Its mission must include enrolling as many members of society as possible in the scientific enterprise, to heighten both their interest in these fields and to encourage them to become players in the game. If we desire to have a public that accepts new technologies and also contributes to their development, we must invite open inquiry and practice full disclosure!

Smart Machines Making Us Smarter

As the *cybergenesis* process unfolds, we will not only integrate intelligent devices into our bodies, but we will go a step further and transplant our intelligence into machines. The concept itself is ingenious—let a machine do the drudge aspects of mathematics, such as calculating, enumerating, and computing, while we attend to higher-level concerns, such as applying mathematics to solving man's problems. Our computers run other machines so that we can busy ourselves deciding what projects we want those machines applied to and how to best use them.

It is quite apparent that the computer, both in its mainframe and personal form, has become one of the triumphs of the modern era, a tool empowering the species to take several quantum leaps forward in its development. The computer enables us to maintain our telecommunications, defense, transportation, and communications systems. Our international banking system would not exist without its infrastructure of supercomputers that execute financial transactions instantaneously and with pinpoint accuracy. In addition, these machines will act as the brain and nerve centers for the spaceships that hurtle us from galaxy to galaxy in our quest to *vitalize* the universe. These machines' efficiency and "intelligence" have tempted some observers to predict that someday they might actually think like humans.

How We Built the Intelligent Machine

In this section we will briefly look at the development of these smart machines, and also consider some predictions that have been made regarding the roles these machines will play in our species' development.

Although the computer as we know it did not appear until this century, the framework for such a device was developed in the early part of the nineteenth century. As he suffered over a table of logarithms one day, Englishman Charles Babbage found himself fantasizing about a machine that could do these dreary calculations for him. Babbage attempted to transform such musings into reality by developing the blueprint for what he referred to as an "analytical engine." Ada Lovelace, Lord Byron's only legitimate child, later joined him in this project, eventually becoming not only Babbage's colleague but also his paramour. As she became more consumed with the project, she provided more and more of the project's brainwork. Many think that her intellectual contributions to the project's final output rival Babbage's. Some who have reviewed her work have labeled her the world's first software engineer—she wrote a paper describing the rudiments of programming, sample programs, and subroutines, all the techniques eventually incorporated in the modern computer. She predicted that future "analytical engines" would play chess and compose music. Although Babbage and Lovelace never actually completed the machine's construction, they provided us concepts such as stored programs, addressable memory, and computer programming itself that would prove invaluable to the development of the modern smart machine.

The first real mechanical breakthrough in computer development occurred in 1940. The English government enlisted mathematical genius Alan Turing to oversee a workgroup composed of mathematicians and electrical engineers to break the code used by the German military in various communiqués. This breakthrough led to the development of the first operational computer, built from telephone relays. When the Germans constructed more complex codes, Turing and his group countered by building better decoding computers. The new machine, called Colossus, was made of 2000 radio tubes. Throughout the war, this new machine, and nine other similar machines running in parallel, were used by the allies to decode German intelligence communiqués detailing military plans and troop movements.

A major boost to computer development came in 1947 when William Shockley and others created the transistor. This new device revolutionized microelectronics and led to the development of an affordable mainframe and minicomputer. Computers made great headway during the 1950s. The public first became aware of the potential of such machines with the use of the imposing UNIVAC on quiz shows and election events. The U.S. Department of Defense quickly came to realize that the computer could be applied to many military tasks, such as predicting ballistic trajectories and simulating battles and whole wars, and began generously funding computer research. Early technological leaps in computer technology led many artificial intelligence (AI) developers to predict that we would soon build machines that could mimic the brain.

As recently as 1960, only a few thousand computers existed in the United States, mostly serving corporations, government, and universities. The personal computer, a consumer-friendly device, did not appear until the mid-1970s. Apple Computer, the first mass-market computer company, was formed in 1976, but it would take several years before the personal computer became a product the average consumer would consider purchasing.[12]

Now, personal computers can be found in over 40 percent of the households in America, becoming part of most people's everyday work experience. They help us communicate with friends and strangers, write our letters, locate information, diagnose diseases, and perform hundreds of other chores. To the extent that they are helping us advance as a species, become more intelligent, and perform more complex levels of calculations and cogitation, we must recognize these devices as an important component of our species' evolution.

A standard yardstick cybernetic scientists use to chart the progress of these machines shows us just how bright our smartest computers are becoming. This measure is called *MIPS*, or a million-instructions-per-second. Each instruction is similar in brainpower to that needed to add two 8-digit numbers. Artificial intelligence and robot programs of 1955, such as the physically immense UNIVAC, had only a fraction of MIPS computing power. This level equates to the brainpower of an insect. Around 1990, real progress was made, and the 1-MIPS level was crossed. Since that time, the brainpower of computing machines, mainframes and PCs, has dramatically increased. The average PC on someone's desk is close to 1000 MIPS, dwarfing the intelligence of the more advanced mainframes of only a few years ago.

We just have to look at the improvement in the smart machine's chess-playing ability to understand how far the computer has come. Chess-playing computers, even the store-bought variety, have always given humans a good game. In 1997, for the first time a chess computer, Deep Blue, actually defeated the reigning human champion, Gary Kasparov. To do this, though, Deep Blue had to be specially wired and supported by over 256 separate computer chips. This computer is a dynamo—it can consider 200 million chess positions a second! When it is playing chess, and only then, Deep Blue possesses one-thirtieth of our estimate for human intelligence, or the full brainpower of a monkey.[13]

The Intelligent Machine Serving Humanity

The computing power of artificial intelligence machinery is growing by leaps and bounds. Even the average home computer owners implicitly recognize that PCs are getting stronger and quicker in their ability to analyze and compute—six months after they purchase such a machine its capabilities are outdistanced by its manufacturer's new line of computers. So we can only guess at how powerful supercomputers, such as the Cray, will become over the next 10 or 20 years.

The Janus computer, developed by the Intel Corporation and installed in 1997 at Sandia National Laboratories, a Department of Energy installation in Albuquerque, New Mexico, has the power to perform a once-inconceivable trillion mathematical operations per second. The Federal Department of Energy's Accelerated Strategic Computing Initiative will use this teraflop machine to simulate the testing of nuclear weapons without incurring the environmental and political risks of exploding them in the real world. Janus is so fast and has so much memory (hundreds of billions of bytes, compared with millions in a typical personal computer) that it can simulate complex events—explosions, nuclear fusion reactions, missile impacts, or the crash of a comet into the Earth—with a fineness of detail and accuracy inconceivable only a short time ago.

Such new technologies are the apotheosis of *cybergenesis*, humanity applying computer technology to promote its own development. Scientists recently used the teraflop computer to simulate the split-second timing of a nuclear fusion experiment. Janus was also employed to enact a scenario in which a comet about six-tenths of a mile in diameter and weighing about a billion tons struck the Earth, vaporizing part of the ocean and setting off tidal waves.

One can only imagine how these new supercomputers can help us understand the workings of the human body and physiology. Once we possess such knowledge we will be well on our way to achieving superlongevity. In addition, artificial intelligence will make a substantive contribution to the species' advancement by broadening our understanding of how the brain functions. In early 2000, the journal *Nature* reported that researchers at Lucent's Bell Labs and the Massachusetts Institute of Technology had developed the world's first computer algorithm that successfully mimicked a key aspect of intelligence, recognizing meaningful patterns in large collections of images, text, or other data. The journal stated that the research provides insights into how the brain's neurons are specifically related to what we would label "intelligent" behaviors. In addition, the computer's masterful capacity for modeling and simulation will enable us to fabricate multiple scenarios of our future physical evolution, including changes in the species' brain structure.

In the next few years, two more teraflop machines at other Energy Department labs will join the Janus computer. The Los Alamos National Laboratory in New Mexico is scheduled to receive a supercomputer built by Silicon Graphics Inc., and the Lawrence Livermore National Laboratory in California will get one built by the International Business Machines Corporation. All three will be connected to form a network of vast computational power.

Scientists hope to speed up computers like Janus a hundred times by 2005 and a thousandfold by 2010 to produce the first petaflop machine. In a single second, such a supercomputer would carry out a quadrillion (a thousand trillion) operations. Such increases mirror the tendency of computer speed and capacity to double every 18–24 months since 1960. Scientists admit, though, that by around 2020 we will have to create new technologies to make our computers run faster, since the key components of computers, transistors, and processors will have become as small as nature will functionally allow. Signals in a computer can travel no faster than light. So to make the computer operate faster, we naturally must move its components ever closer to each other. In addition, the closer the components are, the harder it is to prevent them from frying in their own heat. Other problems in the circuitry will make it even more difficult to increase the speed past levels achieved around 2020.[14]

According to many, if we desire to create a computer faster than the silicon-based petaflop machine, we will have to look to as yet unproven new technologies, such as DNA computing, crystalline computing, and cutting-edge concepts such as quantum computing. According to some, the nervous systems of animals, especially simple ones such as worms, may hold the secret to a whole new direction in construction of computers. Ultimately, the hope is that we will accurately map such organisms' neural activity, translate that activity into programming code, and from there develop a new kind of electronic brain.

Many computer scientists conjecture that machines built on lessons drawn from nature would be so sophisticated that they could accurately forecast the weather for years ahead. Currently we are attempting to achieve such a goal with computers built on conventional technologies. As we entered 2000, the National Center for Atmospheric Research (NCAR) began using one of the world's most powerful supercomputers, an IBM RS/6000, code-named Black Forest, to simulate global climate. With 160 gigabytes of memory and 2.5 terabytes of disk space, it is 5 times larger and 20 times more powerful than Deep Blue, the chess machine that beat Gary Kasparov. Black Forest will help NCAR's scientific divisions and university affiliates conduct research into key weather-related issues, such as ozone depletion, droughts, long-range weather patterns, and global climate changes, as well as humankind's impact on such weather patterns. However, to truly forecast long-term climatic patterns, we will probably need computers using technology more sophisticated than that of the IBM RS/6000.[15]

Scientists think that computer systems that would be built on biological models may not only be more complicated than the ones employed in today's computer systems, but also more powerful and flexible. The memory in these future computers could be derived from simple, natural resources. Pioneering work by Robert Birge at Syracuse University's W. M. Keck Center for Electronics has shown that a light-sensitive protein found in saltwater bacteria could serve as a three-dimensional optical memory to represent computer data.[16] At the University of Sussex, home of the Center for Computational Neuroscience and Robotics, scientists are attempting to develop chips that actually evolve on their own just as organic entities do, becoming more adaptive and powerful as time goes on.[17]

As we overcome the limits of machine computing power, we will discover new uses for the computer and supercomputer. The power of such machines, funded as they are with public monies, should be shared with members of the public who need such massive computing power for their own activities and projects. These supercomputers enable us to perform much scientific experimentation even without having access to working laboratories. Why not, then, make available to members of the public, perhaps *amateur* rocket scientists, geneticists, and computer scientists, the huge power embodied in this convocation of supercomputers, possibly through connections to the home, for their own research and experimentation? Such supercomputers have the potential to transform us all into theoretical and experimental scientists. The layperson, through the phenomenal powers of virtual reality and computer visualization, can learn how science and his universe actually work. Before long, a new non-professional scientific community, the public, will be making its share of scientific breakthroughs.

Robots, Spiritual Machines, and Artificial Intelligence

What role will these smart machines play in the *cybergenesis* process in the near future? Many observers predict that by 2010 computers will assume a major role in education, entertainment, medicine, space, transportation, communications, architecture, manufacturing, and a host of other fields.

One thinker who sees no limits to the computer's capabilities is Ray Kurzweil. This pioneer has always worked on the forefront of computer innovation and invention, in areas such as scanner technology, optical-character and speech-recognition systems, speech synthesis, and digital music synthesis. He holds a degree from MIT and several honorary doctorates. His earlier work, *The Age of Intelligent Machines,* was named the outstanding computer science book for 1990 by the Association of American Publishers. He has founded and spun-off several successful computer technologies companies and is considered an accomplished musician.

In his 1999 book, *The Age of the Spiritual Machine,* Kurzweil predicts stunning developments in computer technology that will have a profound impact on the human species. By 2009, a $1000 personal computer will perform a trillion calculations per second and yet be small enough to be embedded in clothing and jewelry. High-speed wireless communication will provide everyone instant access to the World Wide Web from anywhere. Better yet, we will be able to perform most computer commands verbally, free of the arduous typing still associated with computers. According to Kurzweil, the machines will be getting much friendlier than before: Most routine business transactions, such as making purchases and travel reservations, will take place between a human and a "virtual person" mimicking any number of human personality traits. The Web and computers in general will become an integral part of university training. We will speak to people from other countries through "translating telephones." Pocket-sized reading machines for the blind and visually impaired, as well as listening machines for the deaf will make the disabled more able to participate in humankind's work and contribute to our mission.

Kurzweil's world of 2019 is even better. We will interact with the computer mostly through gestures and natural language. In most transactions, we will be interfacing with a simulated person. Automated driving systems will be installed in most roads. Blind persons will routinely use eyeglass-mounted reading and navigation systems, and deaf persons will read what other people are saying on their lens displays. Paraplegics and quadriplegics will enjoy a whole new world of mobility, routinely walking and climbing stairs through a combination of computer-controlled nerve stimulation and exoskeletal robotic devices. He looks to the day, perhaps 10 years later, in which we will use neural implant technologies to eliminate the limitations we associate with physical disabilities.

He speculates that by 2029 we will have perfected direct neural pathways for high-bandwidth connection to the human brain. Neural implants will enable us to have superior visual and auditory perception, interpretation, memory, and reason. People will be fitted with implants in their eyes that will provide a "high-resolution three-dimensional overlay on the physical world."[18] In addition, what we cannot do with this enhanced brain framework, we will do by interconnecting with the neural net. This *neural net* will be a collection of supercomputers that act in tandem to store information and analyze data, the combination of a global mainframe computer and the Internet. He suggests that we will develop more high-tech methods for individual humans to communicate and interact with this neural net. Permanent or removable implants, kind of like contact lens for the eyes, as well as cochlear implants might be used to provide input and output between the human user and the worldwide computing network. In Kurzweil's future, the possibilities seem limitless. He mentions that while we will not yet be able to download knowledge directly from the neural net to the brain, by 2029 such an achievement will certainly be on the horizon.[19]

Almost the same month Kurzweil's book appeared, a work by Hans Moravec, a robotics and artificial intelligence specialist at Carnegie Mellon, was published that made surprisingly similar prognostications about the enhanced role of the smart machine in the human future. In his book, *Robot: Mere Machine to Transcendent Mind,* Moravec focuses on the application of cybernetic intelligence to "robot-computers." Like Kurzweil, he thinks that we will soon see the emergence of a computer with almost human intelligence, made possible by enabling technologies such as quantum computers (not yet invented), molecular-sized switches, and data bits stored in individual atoms.

Unlike Kurzweil, he thinks that we will want to apply this new cybernetic capability not just to a smarter mainframe. He thinks this new brain power should be used to build a "universal robot" that can be programmed to do almost any chore, such as pick up clutter, open doors, mow lawns, play games, retrieve and deliver articles, guard homes, and eventually run our households. As Moravec says, "Because there is a far greater variety and quantity of physical work to do in the world than paperwork, universal robots are likely to become far more numerous than plain universal computers, as soon as their capabilities and costs warrant."[20]

In Moravec's future, around 2010, two-, four-, and six-legged robots of all sizes will be used to traverse terrain, do heavy lifting, and perform simple grasping of objects. Such robots will have automated vision, be able to make turns on their own, and return to the physical spot from whence they started. We will use these cybernetic helpmates in factories, warehouses, and offices. Moravec even thinks the advanced 2010 model will be able to cook a simple meal, tune up certain types of cars, and perform certain simple assembly jobs. This robot will have the brain power roughly equivalent to a reptile.

Second-generation universal robots will actually have the capacity to learn new skills, by virtue of a processing power 30 times that of the earlier robots. If such a machine makes an error in the performance of a task, it will have the capacity to reprogram itself through trial and error to finally do the task correctly. Second-generation universal robots will probably be used in complicated assembly capacities in factories. Moravec thinks that the speech recognizers built into robots in 2020 should endow robots with the capacity to not only understand speech but detect the emotional intent of the speaker. The computer inside this automaton might be so powerful as to enable the robot to engage in meaningful conversations with humans. He seems to think such ambulatory robots will be used in homes as servants to cook, clean, lift, and move objects. By 2050, robots will be involved in all fields, performing a range of sophisticated activities—they will even assist human physicians in some aspects of surgery. He suggests that by the end of the century, robots, not humans, will design the prototype for the next generation of robot.

These are all examples of *cybergenesis*—machines working in conjunction with the human species to extend and enhance our humanness. They release us from the drudge work necessary to maintain the species so we can tend to more sophisticated productive activities that require higher intellectual faculties. We become better thinkers, more proficient writers, more skilled composers and musicians. Kurzweil even envisions us "jamming" with cybernetic musicians to develop new songs and musical ideas and styles. The machine works with us to help us invent, innovate, and create, becoming a helpmate in our efforts to expand and enhance the development of the universe and ourselves. Moravec's robots will help explore the solar system for sources of power, by mining asteroids and setting up factories in space.[21]

So far, so good! However, in their prognostications both Moravec and Kurzweil engage in questionable leaps of logic. They both predict that their machines will, and should, eventually transcend the human race on *all* levels. Suffice it to say that they both see their inventions achieving a sort of consciousness, a self-awareness that we usually reserve for our conception of human. Kurzweil boldly claims that these future machines will acquire a "spiritual" quality. Both he and Moravec predict that these creations—supercomputers or the "universal robots"—will eventually surpass the human in both the physical and intellectual arenas. The supercomputers will eventually become better decision makers and policy analysts and start to shape the future of the planet and the universe. In other words, the human race will become superfluous and be replaced. Or rather, in their parlance, the human species will "evolve" into these new entities! In Moravec's future, humans throw in their lot with the new machines and decide to become robots. In Kurzweil's, the species downloads its consciousness on to the neural net and becomes part of the "unified being." (All this by 2100, no less!)

Many both inside and outside the artificial intelligence community agree with Moravec and Kurzweil that we should strive to create this machine-human hybridization. Later on I will demonstrate how a widespread acceptance of this vision might impact our economy, society, and political system. I will also present a more optimistic, human-centered alternative to the Moravec-Kurzweil scenario, the emerging expansionary vision of human development and destiny.

Cybergenesis and Human Destiny

The computer, and all calculating and "smart" machines, including the more advanced forms of artificial intelligence, will serve as tools for the human species to achieve its goals. Human progress is the cause for which we created the computer, nothing more, nothing less. And progress we shall have!

The human species will extend and expand its brainpower with a variety of thinking machines. There are no limits to what the human race supported by advanced computer power can achieve. They will enhance our ability to think, create, and perform. They will help us launch our satellites and control their orbits, and help run our energy, financial, telecommunications, and road and air traffic systems. We will even be able to plan and forecast the future of business and society through the use of multiple scenarios drawn up by complex computer programs.

The *cybergenesis* process, the interconnection and interfunctionality of man and machine, benefits the human species in a number of respects. In the most general sense, these machines free us from work such as computation and number crunching that is tedious, repetitive, and more a means to an end than an end in itself. The computer's ability to do this allows us to spend more time involved with more conceptual, abstract activities such as planning, writing, strategizing, goal setting, and higher-level tactical activities. Donald A. Norman, vice-president of research at Apple Computer and professor emeritus at the University of California, San Diego, claims that the human and computer make perfect work partners. Since the computer is more accurate than he could ever hope to be, Norman is more than happy to allow the computer do his calculations for him. As Norman puts it: "I think about the problems and the method of attack; it (the computer) does the dull, dreary details of arithmetic or, in more advanced machines, of algebraic manipulations and integration." Norman, a psychologist by trade, claims that this synergy allows man and machine to form a "more powerful team."[22]

Another benefit of the computer age is the development of a "neural net," somewhat along the lines prophesized by Kurzweil and most cybernetic seers.

The Internet is a repository of information, expertise, opinion, and computational and analytic power never before witnessed in human history. The Internet is becoming indispensable in so many areas of human activity that eventually people will literally not know how to live without it.

Nevertheless, for such a neural net to be useful it must be routinely and permanently accessible to each member of the species. The question is how do we make this computational power and information more available, even immediately accessible to human beings for their use and benefit? Kurzweil and others suggest that humans will eventually connect themselves either permanently or semipermanently to a mainframe-like neural net. British Telecom is supposedly working on fantastical technology that could presumably connect brain and computer. If this came to pass, the user of such technology would be like the person spending countless hours in front of the PC surfing the Net, except in this scenario he would be permanently, and directly, in touch with the Net, while awake and asleep. In other words, humans will literally become part of this network. There is an ever-growing, evolving intelligence, but it won't be yours or mine. This ever-growing intelligence will increasingly become the sole property of the machine—with you and I as mere participants in this group mind.

Without debating the moral or ethical implications of such a Net at this point, let me just say that this system as proposed is both impractical and counter to humanity's efforts to achieve its destiny. We need a system that is friendly to geographic mobility. That is, as a spacefaring society, we must be able to bring this intelligence with us. In short, we need a system in which we don't join the neural net, but rather it joins us.

I am suggesting that we will discover ways to get that valuable information accumulating on the Net back into the human brain and allow the individual to digest and utilize this data and information at almost supra-human levels. The externalization of all this stored information on the World Wide Web, far from being an end-point as envisioned by Kurzweil, is a temporary phase. We will invent methods to bring this information into the brain, to make the brain function more efficiently. Within 20 years, the goal, if not the very definition, of cybergenesis will be transforming what is external to the human being into an internal property.

I envision individuals owning units, such as battery-powered intelligence packs tuned to the Net, which they will use to download information and upload queries and computational problems. The communication between the chip-implanted human and the neural net will take place via these "proto-pac" intelligence machines that individuals will carry with them or have in their homes. *With nanotech, small-chip technology, we can bring whatever we develop out there, all that information and computational power, inside us.* We will communicate, download, and upload, at will. Our will! In this way, whenever we desire we will be able to gather and utilize information from a neural net that calculates at a rate of a trillion bits per second. Bear in mind, though, that our relationship with the neural net will always be a temporary one. Because human existence involves much more than calculating, computing, and manipulating information, we do not need to be constantly connected to the neural net.

Another development I foresee will involve the combination of biogenesis and cybergenesis. I believe that once we possess a more complete understanding of how the brain operates, we will develop ways to enhance human intelligence by modifying brain structure. This may involve a combination of permanent computer-based modules placed in the brain and a variety of pharmaceutical, genetic, and laser-based surgical solutions. It is hardly a stretch to imagine molecular-size computers and nanocomputers inside the human brain enhancing our memory, thinking ability, perhaps even our communication and visualization skills.

We will increasingly turn to the computer to expand our comprehension of how large systems operate. Future computers will excel as a simulator and modeler of various systems and subsystems, including the universe, the human body, the global economy, financial markets, global climate, and matter itself. Soon we will further research and gain an understanding of our world by using technologies that combine virtual reality with these highly detailed visual simulations of systems. Imagine using this hybrid-technology to become a "shrinking scientist," who can wander around atomic structures and "feel" the forces combining the nucleus and electron by merely putting out our hands and "touching" quarks and neutrinos. We could train for expeditions to Venus or Mars in virtual reality before we even made the voyage, acclimating ourselves to obstacles we could encounter on our voyage. We could also use such experiential models to research the evolution of the universe or the behavior of viruses in the body. In this way, the computer becomes the ultimate research tool.

The ability of such technologies to mimic reality is both their strength and their weakness. Eventually the computer's increased power will enable it to realistically simulate any environment or experience—a day at the beach, or an afternoon with a virtual version of your favorite actor or actress. This use of simulations for entertainment purposes seems harmless enough. However, let us recognize that such simulated virtual paradises could become addictive.

Even supporters of this technology, such as Kurzweil and researchers at British Telecom, have warned that some people might come to prefer these simulations to reality itself. Simulation technology then becomes mere mind candy, an opiate that muddles and distorts rather than clarifying and elucidating. If people begin mistaking virtual reality for reality itself, progress, both human and machine, would stop dead in its tracks.

One novel area in which computers are making their mark is as an aid to the nurturing and facilitation of human creativity. In a true example of cybergenesis, the computer is making musical composition as easy as word processing. One can write a song, enter it on the computer, hear it played back by electronically created instruments, make changes, hear it in different keys, and even publish it online. In addition, you would never have to hire one musician or even pick up a musical instrument to complete this process. The ease of composition and musical production permits the composer to focus on his or her creative contribution and not worry about the small details. Today nearly all commercial music, movie and television soundtracks, and recordings are created on computer music workstations. These programs can synthesize and process sounds, record and manipulate the note sequences, generate rhythm and bass lines, and help the composer develop melodic progressions and variations. Now all of us can compose symphonies and conduct entire orchestras if we so desire.

Computer programs exist that help us design a home, invent new products, and learn a plethora of subjects. This is all part of the cybergenesis process, the machine enabling the human being to become more original and more brilliant.

The computer has another benefit, albeit an unintended one. Computers, when programmed and wired to perform only one task—statistical computation, disease diagnosis, traffic control, or chess playing—often function better than human's ordinarily do at that particular task. *By performing and thinking on such a high level, they set a new benchmark for human performance in all areas.* This is a much misunderstood and often overlooked aspect of the human-computer interface.

In 1996, Gary Kasparov, reigning world chess champion, defeated the best computerized chess program/computer in the world, IBM's Deep Thought. As mentioned earlier, Kasparov's 1997 rematch with the new IBM program called "Deep Blue" did not go as well for Kasparov. After winning the first game, Kasparov drew the next three, and then started to lose, consistently and badly. To the shock of the chess world, Kasparov eventually resigned the match.

The artificial intelligence crowd made much of IBM's defeat of the Grand-master many considered the greatest chess player in history. Some felt that Kasparov's defeat "proved" that the artificial intelligence machine had now surpassed humans at chess playing. But wait! People ignore the fact that Kasparov, by playing the best player in the world, albeit a machine currently playing better than any *human* opponent Kasparov has ever met, provides Kasparov the opportunity to reach a higher level of chess-playing proficiency. In fact, he can play the next match against Deep Blue's successor at a level he could not previously have envisioned, since he had never played any hu-man as talented as Deep Blue.

Therefore, computers not only act smarter in some fields, but they have the ability, as in the Kasparov example, of making the human smarter, more proficient in all skills. The computer shows us new approaches and teaches us new strategies we can use to reach ever higher levels of skill. I call this the "cybernetic benchmark effect," the process by which the computer, by hold-ing us to ever-higher performance levels, forces us to rewire our brains and reconfigure our cognitive processes in order to operate at totally new levels. The computer transcends its role as teacher to become a competitor that forces us to run faster, intellectually speaking.

Cybergenesis is changing our world for the better. The smart machines that permeate our society will help us reach new heights. It provides us more information than we ever dreamed possible. Computer-based implants might improve our ability to hear, see, and think more efficiently. In addition, to quote Ray Kurzweil, "the creation of new technology is fueling the expansion of economic well-being."

However, progress has its costs. The cybernetic and robotic revolutions eliminate jobs at the bottom of the skill ladder as they take over much of the drudge work of computation, manufacturing, mining, and assembly. Robots do these jobs better and quicker, and unlike humans, never fatigue. So hu-mans must become engineers, managers, planners, and analysts. Those who don't upgrade their skill levels to fill such positions will quickly sink to the lower levels of the economic ladder. Millions of workers are displaced and unemployed, even while skilled jobs such as engineer and software program-mer go begging.

Technology also deprives us of excuses we might be tempted to devise for not succeeding in these higher-level jobs, since it provides us the tools to do these difficult jobs better. The average manager now has at his beck and call sophisticated computer systems to help him communicate, write, and develop proposals. The manager can no longer blame his incomplete reports on lack of access to information—the Internet provides all the information he needs. Because of the computer, salespeople and marketing managers are expected to make sales forecasts using statistical techniques that were once solely the tools of Ph.D.s in the social and physical sciences. All of us, professors, lab technicians, business managers, are expected to learn the skills necessary to incorporate the computer into our jobs and our lives. Moreover, we are expected to show improvements in performance that these tools make possible.

One of the dangers related to cybergenesis is one endemic to progress in general. As we incorporate cybernetic technology into our lives, and our bodies, we do tend to forget that, as smart as computers are, in the end they are only smart *machines*. They perform some tasks so well we imagine they can do almost anything. Because they calculate so quickly and accurately, and present information so completely and readily, we assume they can also think, create, plan, and set goals. Such assumptions lead to a false sense of confidence in and excessive reliance on the machines.

To understand the machine's limits, let us look at the interfunctional relationship of man and computer in the music area. While composers and artists incorporate the computer into many aspects of the creation process, few would suggest that these machines act creatively, as the term is commonly understood. Some computer programs enable the computer to supposedly "create" music on its own. However, such composition is entirely based on human musical input and programming ability. One such program, called EMI, is an ongoing project of composer-computer maven David Cope. EMI distills essential patterns from bodies of work by other composers and produces new music in these artists' styles. In recent years, EMI's classical compositions have reportedly pleased audiences, who in blind tests rate EMI's works above that of many human composers. Writers like Hans Moravec like to claim that the EMI case demonstrates that computers can learn to write music.[23]

In truth, this machine is not inventing or writing music in the real sense. It distills the music Bach, Mozart, Beethoven, and others have composed and then synthesizes and restyles the work. It is doubtful that fans of EMI's musical throughput would find the computer's music so pleasing were it not based on Bach's original themes and musical inventions.

Will robots ever write totally original music humans care to hear? In spite of the fact that programs such as *Improvisor*, written by British jazz saxophonist Paul Hodgson, can emulate styles ranging from Bach to Louis Armstrong, machines will not "compose" the world's next great symphony.

From the viewpoint of improving man's consciousness, this debate is a minor one, anyway. We are concerned with cybergenesis, the use of the smart machine to improve the human condition and human performance. The ability of the computer and music programs to democratize musical composition ensures that the twenty-first century's next Beethoven will find it easier to compose her or his music. The computer contributes to cybergenesis by facilitating the creative process.

Even Deep Blue's greatest fans know where to draw the line. Those who watch Deep Blue play chess feel that the program does more than merely calculate its next move. It does exhibit a sense of strategy and a feel for the ebb and flow of the game. However, no one dares use words like "intuition" when describing the machine's game style.

So let us maintain our focus on the true meaning of the computers, robots, and the like—these technologies make the human species smarter and our life better and more enjoyable. They are primarily a complement to human activity—nothing more, nothing less. That will be their role even if they achieve the level of cybernetic sophistication characterized by the android Data on *Star Trek: The Next Generation.* Increasingly, computers will help managers, CEOs, economists, and policy analysts to create multiple scenarios of future events, providing them with a rich array of alternatives to consider when making decisions. It will still be the responsibility of humans to perform higher level processes, including the planning, development, and creation of organizational mission statements and visions for the species. And it will certainly be humans who create the next generation of robots and smart machines.

These ever-smarter machines will enhance humankind's ability to spread our humanness and consciousness throughout the cosmos. Cybergenesis will provide us with the brainpower and mental dexterity to perform the computational and conceptual feats required to vitalize distant spheres. Computer-based enhancements to the body will make us sufficiently durable to survive long journeys into space. The supercomputer will help us understand our universe better so as to aid in its evolution and progress.

Most importantly, though, are the subtle changes that will take place in the human species itself as cybergenesis, the partnering of human and machine, proceeds. By the very activity of inventing ever-smarter machines we create a much smarter human. We send our humanness, our intelligence, into a machine that can think at increasingly higher and complex levels. It performs, for instance, seemingly impossible mathematical and computational feats of magic. We, then, reintegrate the products of this intelligence into our culture, our society, and eventually ourselves. We are becoming superior humans as a result of this constant interaction with our own invention, the smart machine.

In short, the humanness, the intelligence we are exporting to and extending across the universe is far superior to that which we would be sending had cybergenesis never transpired. Cybergenesis not only makes it possible for humankind to effect vitalization; it helps improve the final product that we are to export, human consciousness.

4

Species Coalescence

People commonly remark that "the world is shrinking," or state that "the globe is getting smaller." Such statements are usually sparked by any number of trends: we now enjoy instant communication with anyone across the globe; transportation advances have made it easier for the average person to travel anywhere in a relatively quick time frame; and many of us work for organizations which are themselves global in nature. People and places that used to seem distant now appear extraordinarily close.

While such trends have made our lives richer and our work experiences more intense, the shrinking of our globe is having a more profound impact. We are being transformed from residents of our community, town, or nation into members of a much larger entity, a global society.

I label this process *species coalescence*, the process by which humanity increasingly interconnects in a multitude of ways. Species coalescence is the fourth force, along with processes such as *biogenesis, cybergenesis,* and *dominionization*, hastening the human species to evolve toward a higher state. By achieving the degree of intense unity implied by the term *species coalescence*, our species will not only continue to expand its material and social progress; it will be in position to reach that loftier set of goals I refer to as *vitalization*.

As we begin the twenty-first century, several major technological revolutions are accelerating species coalescence. Humankind will more easily connect physically by virtue of the emerging global transportation grid. Moreover, we will enjoy a closer meeting of the minds, an intellectual coalescence via a highly sophisticated planetary communications network. In addition, an emerging web of human economic activities, which I label the *universal production/consumption system*, will generate an economic symbiosis that will cultivate in all of us a heightened sense of identification with other members of our species across geographic and cultural borders. Moreover, those aforementioned communications and transportation breakthroughs will make it easier for all of us to contribute our diverse skills and talents to this universal production/consumption system.

Throughout this chapter, we will examine those breakthroughs in technology empowering humanity to reach this level of global unity unimaginable only a century ago. We also will come to see how the gradual coalescing of the human species will make us more capable of attaining what I consider our overwhelming destiny, the vitalization of the earth and eventually the universe.

The Global Transportation Grid

For species coalescence to occur, we must be able to physically interact with each other quickly and easily. The easier it is for a person to viscerally experience others' countries and cultures, the more likely it is that he or she will feel solidarity and unity with people from all over the world. Transportation technology over the last century has changed the way we live. The airplane, the automobile, and other inventions have shrunk distances between cities and nations. As you will see, daunting and mind-boggling breakthroughs will soon make Europeans, Americans, and Asians quite accessible to each other. In the next millennium, we will all be neighbors!

The Highways and Cities that Will Connect Us All

It took nearly two centuries to plan, design, and build the Chunnel, which now connects England and the European mainland. Now, nations and corporations are developing a plethora of under- and over-ground connections—bridges, tunnels, and causeways—that within the next two or three decades will link most of the globe's landmasses. This network will comprise a veritable "global superhighway" serving over 100 nations on five continents.

The planners of this global network ask us to imagine driving a car from Scandinavia down to Europe; across Germany, France, and Spain; and then through the "Gibraltar Tunnel" to Africa. You would then circle the African continent, leave the Eastern Mediterranean, and drive across Asia and China before heading up the Pacific Rim to the Arctic. Proceed along the "Bering Tunnel" connecting Russia and Alaska, and join the Pan-American highway across Canada, the United States, down to Mexico, and Central America.

You will be able to easily travel from Central America to Colombia, the northernmost country in South America, because you drove over an elevated highway that spans the Darien Gap. You then continue your trip to the southern tip of South America.

The planned projects for causeways and tunnels will make such a trip a reality fairly soon. The Euro tunnel, the Chunnel, has already been built and now serves as a major intercontinental thoroughfare. A series of bridges connecting Sweden to Denmark to Germany is currently under construction. A "Gibraltar Crossing" linking Europe and Africa is in the planning stage. Several proposed African routes await financial backing. Submarine engineer J. V. Harrington has developed a plan to connect Europe and North America via the North Atlantic. His plan would involve long tunnels connecting Labrador to Greenland, Greenland to Iceland, and Iceland to Scotland and Norway. Surface transportation would extend across both Greenland and Iceland. Such a route would be suitable for car and rail.[1]

Possibly the most ambitious plan of all is an Asia-Europe connection, the so-called new Silk Road. Various routes are being considered: a southern-based route stretching from the Mediterranean to Singapore; a middle route from Turkey around the Black Sea, across China to Shanghai; and a northern route running from Berlin to Moscow across Siberia and Mongolia to Beijing. This scheme calls for a tunnel running from Siberia to Alaska, construction of which could begin in a few decades, once political, engineering, and environmental issues are settled.[2]

The shape and form of the city of the future will also accelerate the species coalescence process. There will soon emerge the *supermetro*, a city-like entity spread over a very large geographic area but completely functionally integrated. According to one analysis, the supermetro will extend out perhaps 100 miles in all directions from its *metro center*, which will contain urban services such as an international airport, medical centers, convention centers, universities, and a domed stadium. A *satellite area* will contain high-performance business parks and office and/or industrial areas. A *hinterland* will be situated near shores, lakes, or mountains. Quiet and secure residential areas will feature shopping and good schools. Planners are defining a super-metro area as one in which a person could make a round-trip journey, a commute if you will, by high-speed surface transport within a typical business day. Over the next few decades, supermetros might arise that span the entire Boston-to-Washington corridor, much of northern Europe, and most of Honshu, Japan's main island.[3]

Highway systems servicing global and local travel will most likely incorporate next-generation, intelligent vehicle and roadway technology that will enable us to safely travel at extremely high speeds. These new highways will be specially equipped to provide their users navigation guidance, vision enhancement, and cruise control to ensure driver comfort and safety. To make personalized travel at extremely high speeds a reality, we will drive "smart cars," equipped with computer chips that can send to and receive a great deal of information from sensors situated on roadways. Drivers will pay tolls without stopping and have access to a wealth of instant data to help them monitor local highway traffic.

Brookhaven National Laboratory on Long Island is developing a smart road system featuring cars that could travel at lightning speeds, 200 miles per hour, guided robotically by a "precise positioning system." These extremely lightweight single cars could accommodate as many as a dozen people. The Brookhaven system would serve both local traffic as well as cross-continental travel, a 16,000-mile network of roads crossing the United States and linking many major cities. While the system's $200 billion price tag might seem high, it represents only about 1 percent of what we plan to spend on transportation over the next 20–30 years. It would link small and large cities, since one could construct a system, which could stop almost anywhere—towns, villages, shopping malls.[4] In addition, it would generate revenue!

It may seem odd that a book about vitalizing the universe accords so much importance to land-based travel. After all, vitalization ultimately requires that we find a way to travel quickly and easily from one sphere to another. So one might expect that we should be more concerned about advances in aerospace and rocketry, such as hypersonic flight, that will make it possible for us to traverse the universe. Certainly, innovations in air travel that we will describe forthwith will literally obliterate distances around the globe and shrink the world for us.

However, species coalescence entails more than just getting from one place to another quickly. It requires that the individual understands and shares the experiences and perspectives of others. Ultimately, the truly global society would be one in which all members can easily exchange ideas and viewpoints with others, experience the various aspects of the "other individual," including his or her physical locale, and be able to integrate themselves into that person's society, and he or she into yours.

So the question before us now is "Which form of transport will bring mankind together most completely?" Surely, land travel promotes a rich and profound interactive experience between traveler and destination. As we travel by train, or the supercars envisioned at Brookhaven, we experience—see, hear, even touch—the areas, cities, farms, villages, we are traveling through. We may frequent local stores and talk to the local people to get a sense of how the land, cities, and people fit together. The level of interaction is unparalleled. In addition, just extend this type of experience over thousands of miles and a number of weeks if one were to drive the "global superhighway." As you drove from Florida to Paris to Egypt to China, you would gradually acquire a sense of the inherent connectedness of the species as well as your place in this wider picture.

The Advent of High-Speed Rail

All forms of travel have their distinct advantages. Air travel can get us most places quickly. Global highways and intelligent cars will most certainly bring us closer together, but not at supersonic speeds. Wouldn't it be nice if a technology existed to combine the advantages of land travel with the speed of air travel? Breakthroughs in rail transportation might deliver to us the best of both worlds!

We have made tremendous progress in the development of high-speed trains over the last few years and will make even greater strides in the first part of the twenty-first century. The Japanese Model 500 Nozomi and France's TGV travel upward of 160 to 180 miles per hour. However, the proposed *maglev*, or magnetically levitated trains, will far exceed these trains' speeds. Maglev will run on basic magnetic principles: When two magnets are put together opposite poles attract, similar ones repel—the train moves because it is simultaneously being pushed and pulled by a series of magnetic poles placed closely together all along the track. In both cases, the train moves as the magnetic field travels along the guideway. One reason that trains propelled by maglev systems will run so quickly is because the electromagnets to be used in the process will actually lift the train 1 to 4 inches above ground. Therefore, these trains will operate entirely free of the friction caused by wheels running on track that traditional trains endure.[5]

How fast can such trains travel? In spring 1997, the Japanese initiated an ambitious program to develop magnetically levitating trains that will travel at speeds greater than 300 mph. The proposed route linking Tokyo and Osaka by maglev train would reduce travel time between the two cities to under 1 hour from the current 2½ hours. By fall 1997, the Japanese three-car MLX01 set a world speed record of 279 mph for a manned maglev train. Many feel that this train may soon exceed the incredible speed of 340 mph.[6]

France will connect its TGV system to Belgium and then to the Nether-lands, and extend it westward to Barcelona, Spain. Russia has announced plans for a high-speed link between Moscow and St. Petersburg, while Italy forges ahead constructing train lines that connect Turin to Venice and Naples to Milan. Plans are in the offing to construct high-speed links between Cas-ablanca and Cairo, Quebec City and Toronto, and Hamburg and Berlin. Switzerland plans to link its major cities with a 250-mph underground mag-lev rail system called Swissmetro.[7]

The United States finds itself at a crossroads in its mission to install high-speed rail throughout the country. Amtrak plans to begin operating 150-mph trains between Boston, New York, and Washington by 2001 or 2002. In early 1999, Transportation Secretary Rodney Slater announced that Am-trak and the nine states of the Midwest Regional Rail Initiative would invest millions of dollars to prepare high-speed rail corridors from Chicago to In-dianapolis to Cincinnati. This ambitious plan would provide trains running as fast as 120 mph to states such as Illinois, Michigan, Wisconsin, Minnesota, Ohio, and Nebraska.[8] Congress has authorized that over $1 billion be spent to begin upgrading the U.S. rail system to a high-speed transport network. Congress realizes the advantage of using technologies that could reduce the Pittsburgh to Philadelphia voyage from 8 hours to 2½ hours and the Balti-more to Washington trip from 1 hour to 15 minutes.

Obstacles remain, however. In 1999, newly elected governor of Florida Jeb Bush killed a visionary project linking Orlando, Tampa, and Miami with French-style TGV trains running close to 200 mph. Bush questioned the projected ridership and whether the routes might lead to unwelcome development around the Everglades. He claimed that he could not see Florida enjoying sufficient economic return on such an investment.[9] Mem-bers of pro-high-speed rail lobbying groups recently revealed to me that resistance to high-speed rail comes not from the general public but from Congresspersons, airlines, and environmentalists. In one sense, these new high-speed train systems promise to be too successful as people movers. When France opened its first TGV line between Paris and Lyon in the late 1980s, the internal French airline, Air Inter, had to cut back on its service on many local air routes. Rumor has it that Southwest Airlines actively campaigned against the construction of a Dallas-Houston high-speed train link. The slow pace of adoption of high-speed rail in the United States is especially ironic in light of the fact that two Americans initiated the maglev concept. Although these Yanks were granted a patent on a maglev train design in 1968, suspension of federal research funding in

1975 permitted Germany and Japan to lurch into the lead in development of the critical technology.[10]

Finally, in May 1999, the high-speed rail enterprise got a welcome boost from the federal government. Transportation Secretary Rodney Slater announced that Baltimore, Pittsburgh, Atlanta, and other cities would share $12 million to study the feasibility of building maglev style trains that could travel up to 310 mph. The three lines that are getting the go-ahead would link Baltimore and Washington, Pittsburgh's airport with the city itself, and a corridor between Atlanta and Chattanooga, Tennessee.[11]

Meanwhile, countries around the globe are adopting this technology at an extraordinary pace. The Anglo-French Company GEC-Alsthom has contracted with South Korea's High-Speed Rail Construction Authority to help South Korea build its high-speed rail system. The first line will run between Seoul and other Korean cities. To gain entry into the market, the French company had to agree to permit Korea to keep some of the French technology. By having this advanced technology, Korea can become a major player in designing and building high-speed rail systems throughout Asia. This growing Asian power is now planning to construct high-speed lines between South and North Korea, as well as throughout China.[12]

Traveling at speeds exceeding 300 mph, the supertrain will link cities, people, jobs, countries, and cultures. We can only imagine how this technology will accelerate the coalescing of the human species, as these trains travel along new global superhighways and through intercontinental tunnels that connect Europe, America, and Asia.

Turnpikes in the Sky

Within this decade we might witness the emergence of a whole new method of travel, one that combines the ease and speed of air travel with the privacy and independence of the automobile. Moller International, a California-based aerospace company, has developed the Skycar, a four-passenger sedan that can streak through the sky at 350 mph. Its inventor, Paul Moller, claims that the Skycar, scheduled to begin its first trial flights in mid- to late 2000, will eventually replace the automobile. Moller advises us to "forget the automobile . . . the automobile is headed south."[13] NASA believes that Moller is on to something. According to Dennis Bushnell, chief scientist at NASA's Langley Research Center, the market and the technology are already in place to make these individual skycars a reality, barring unnecessary intrusions from government regulatory agencies.

Those who have actually seen the Skycar prototype liken it to a modified "Batmobile"—it has a tapered body and a regal single wing rising like a gigantic spoiler off the tail. The passenger compartment is covered with a bubble-like glass canopy that will allow its four passengers to get a good view of the landscape and the sky above. The current model has a range of 900 miles and gets about 15 mpg using regular automotive fuel, roughly equivalent to the fuel efficiency of today's SUVs. A revamped Wankel engine, whose technology Paul Moller purchased in 1985, powers the vehicle. The Wankel is an extremely efficient and powerful engine whose light weight enables the Skycar to lift off and soar with little effort.

Many of Moller's supporters claim that Skycar will completely change the way people live. The conceptual differences between flying and "driving" will vanish, since this car will operate both in the skyways and on the highways. And if a driver does not trust his navigating ability at 20,000 feet altitudes, he can choose to "fly by wire" by turning the car's operation over to a computerized satellite tracking system. According to NASA's Bushnell, in a world in which we all travel on computer-controlled Skycars moving efficiently through the skies, we would no longer have to sit in highway traffic jams.

This high-speed Skycar will accelerate the trend towards species coalescence. With the availability of 350 mph Skycars, we could safely and easily travel anyplace we want in a short period of time. We could also live anywhere we desire. Bushnell sees whole new communities growing in previously desolate areas. "There is a lot of empty space in this country," Bushnell said. "With telecommuting, and this machine, you can live wherever you want."[14]

Hypersonic Flight—The New Burst of Speed

When we finally develop the technology to transport anyone from one part of the globe to another in a few hours' time, places that in earlier times I might have considered a foreign country will seem more like an adjacent city, town, or even suburb. In addition, we will be more likely to consider that country's denizens neighbors rather than foreigners. After all, how culturally or socially distant can you feel from someone when you can travel to his or her geographic locale in a few hours time at most?

Engineers at NASA, Boeing, General Electric, and Pratt and Whitney, with some input from the U.S. government, are working on a High-Speed Civil Transport (HSCT) that can travel at incredible speeds. This craft will carry 300 passengers from Los Angeles to Tokyo at Mach 2.4, 1600 mph, three times faster than today's jets, cutting the flight between those two cities down to under 5 hours. Traveling from Tokyo to Los Angeles would take 4 hours, Sydney to Los Angeles 7 hours, with a 1-hour stop included. The HSCT is being engineered to be quiet and environmentally benign. It will measure 320 feet from nose to tail, 90 feet longer than a 747 and slightly wider than a Boeing 737. Because of engineering improvements, the HSCT will enjoy a 5700-mile range, almost twice that of the Concorde.

The HSCT should improve on the Concorde in all phases of air travel. After all these years of service, the much-heralded Concorde still flies only three routes, all between the United States and Europe—the price premium associated with this aircraft can be supported only in these three markets.[15] The high cost of and limited access to the Concorde has led some to dub it the "fat cats' express"—supersonic speed for one class of people, subsonic speed for another. Mass production of the HSCT by 2010 should democratize high-speed travel and provide all humankind with the superfast global transport it needs.

However, the speedy HSCT pales in comparison to the potential velocity of hypersonic vehicles that might be available in the foreseeable future. Hypersonic flight that will transport people at incredibly high speeds will become a reality when various aerospace agencies and companies complete the development of the space plane. While most of the research on hypersonic flight is aimed at building a better space shuttle, eventually this research will be applied to civilian transcontinental flight. Soon, planes will streak across the skies at ultra-high speeds, revolutionizing the economic and social relationships between nations and people throughout the world.

NASA is scheduled to produce its hypersonic X-33 by 2000 or thereabouts. This experimental aerospace vehicle, which takes off like a conventional rocket, straight up, but lands on the runway like any other airplane, can reach speeds of upwards of 4000 mph traveling at altitudes of 50 miles or more. It will be able to cross the American continent in 30 minutes. On longer journeys, the plane will reach speeds of Mach 15, or 11,000 mph. According to Dan Dumbacher, NASA X-33 deputy program manager, "By developing and proving these systems, we're creating the ability to build space planes that eventually will fly to orbit, return for servicing, and launch again as often as today's commercial airplanes make scheduled flights."[16] The X-33 is a prototype for a much larger reusable launch vehicle called VentureStar, which its developer, Lockheed Martin, insists it can rollout by 2004. VentureStar will serve a great many purposes, such as launching satellites and monitoring global agricultural and environmental conditions on Earth. The monetary returns on such applications of the VentureStar will ensure that this craft will be profitable.[17]

The Universal Production/Consumption System

Transportation shrinks the time it takes for people to physically connect with each other, and as we shall see, advanced communication technologies enable us to transmit verbally and visually sound and images. Both technologies facilitate interaction between whole groups of people and make each of us more familiar with others.

However, people come together in more subtle ways than those two modalities. The rapidly expanding universal production/consumption system is quietly nurturing the species coalescence process. A certain bond develops among people who work together in cooperative ventures, be it a team, a work-group, or a multinational corporation. Moreover, they also come together when they develop a sense of interdependence born of consumption behaviors and patterns. The more our needs are met by others, the greater our sense of allegiance to them. Over the last several decades, this interdependence has grown dramatically as nations and regions come to depend on other societies for their foods, furniture, cars, electronics, and textiles.

In the emerging Macroindustrial Era, the world population is being integrated into a system that provides people access to all products and services and opens opportunities for them to contribute to the worldwide industrial production process. The new system permits all members of the species to participate in the production process—different nations can become involved in the research and development, design, manufacturing, and marketing of a single product. The assimilation of the entire world population into this process will help us reach true species coalescence.

Over the last several centuries, trade, commerce, and production have become increasingly more global in scope. Yet for most of this period a tremendous chasm has existed separating producers and consumers, owners and workers—the industrial haves and the have-nots. The global economy more resembled colonialism in which one group of countries did the inventing, the manufacturing, and the distributing of the world's goods and services, and another group served merely as a source of raw materials and cheap unskilled labor.

Now, the world economy is evolving into a system where all countries participate as both producers and consumers. Former third world countries that provided only raw materials and labor to the global market now participate in the global economy as producers, innovators, workers, and consumers.

Coalescence is accelerated by the interdependence of nations—if country A needs country B's goods and services, and vice versa, they will work together closely, become more intertwined, and become alike on many levels. Developing countries have a commodity that the rest of the world especially needs: a large and increasingly well-educated workforce. Countries like China, India, Indonesia, Brazil, Pakistan, Thailand, Mexico, and South Korea will provide the world its producers. The worldwide workforce has grown dramatically, from 1.5 billion in 1970 to an estimated 2.7 billion in 2000. However, most of that growth comes from the aforementioned countries. These countries have a far younger workforce than the United States, Japan, Germany, and the United Kingdom. (A 1999 cover article in *U.S. News and World Report* claimed that the West was facing an "aging population crisis.")[18]

To be needed, nations as well as individuals must produce at world-class levels. Jeffrey Garten, Dean of the Yale School of Management who also served in the Clinton administration as U.S. Undersecretary of Commerce for International Trade, thinks that during the early years of the twenty-first century at least 10 nations will achieve world-class status. The list shows how quickly the universal production/consumption system is evolving: India, China, South Africa, Mexico, Argentina, Brazil, Poland, Turkey, Indonesia, and South Korea will all soon profoundly impact world trade, global financial stability, and the transition to free-market economies of all the countries of Asia, Central Europe, and Latin America. Moreover, all of them have the ability to profoundly impact Western interests.[19]

These nations' recent economic indicators demonstrate just how much a part of the universal system each is becoming. Between 1991 and 1995, China's gross domestic product increased an average of 11 percent per year. South Korea, Indonesia, and India each averaged around a 7 percent growth rate per year. Taiwan, Hong Kong, Singapore, Thailand, Malaysia, and the Philippines also enjoyed high growth rates. By contrast, American and European growth rates did not rise above the 2 percent rate until the late 1990s.[20]

Even though these Asian countries were caught in a slide that lasted a year or longer due to currency fluctuations and crises, they still showed signs of recovering nicely by 2000. The Indian government announced that its GDP level was again moving up. Other countries that had been hard hit by the 1997–98 financial madness by 1999 were back to announcing new deals for power plants and car production. Malaysia's auto industry was coming back quickly—new motor vehicles for 1999 would be 12 percent higher than 1998.[21]

These countries are integrating into the system not just by producing goods, but by expanding their capability to consume others' products. China is buying 10 percent of all Boeing's planes. India and Singapore are also major purchasers of aircraft. Many of these countries are providing the American defense industry a much-needed infusion of money. (When in 1998 the shaky world economy forced some of these countries to suspend purchases of aircraft, companies like Boeing immediately started to lay off workers.) These countries are building super-modern transportation and energy infrastructures, often with the help of American and European companies. In addition, the developing countries are becoming mass importers of consumer goods such as soft drinks and entertainment products such as movies and CDs.

The International Financial System: A Necessary Nuisance

One way the universal production/consumption system accelerates species coalescence is through the development of a method for moving money and investment into vital infrastructure and consumer product projects around the globe. The speed at which this process operates simultaneously strikes fear, admiration, but mostly awe in all who witness the system at work.

For better or worse, we are all part of and subject to the international financial system. For instance, European governments have been heavily investing in South American private-sector businesses such as banking and food throughout the 1990s. Europe has been successful in establishing trading agreements with Mercosur, a South American NAFTA-style trading bloc that links Brazil to Paraguay, Uruguay, Bolivia, Argentina, and Chile. In Brazil, South America's biggest market, 7 of the 10 largest private companies are European-owned. (Only two are American.) Europeans dominate Brazil's auto industry (Volkswagen and Fiat), supermarkets (France's Carrefour chain), and personal care products market. Spain's Telefonica de Espana has spent billions buying telephone companies in Peru.[22] A Canadian investment consortium, The Royal Group of Canada, is planning to finance manufacturing plants in the Philippines.[23]

Europe and America are being challenged on the global economic chessboard, often by many former third world countries. Japan's companies set up automobile manufacturing ventures in China; South Korean companies establish paper production plants in China; and Indonesian firms drill for oil in a variety of Far Eastern countries. A Malaysian company constructed and now manages a hydroelectric plant in China's Yunnan province, while NEC and Mitsubishi Electric offered Thailand support in launching three observation satellites between 1999 to 2002. The two electronics giants will design and develop Thailand's satellites, sensors, and experimental equipment.[24]

China will play an ever-larger role in international and domestic investment circles. China signed a series of multibillion-dollar foreign deals that helped it gain access to energy and minerals in Central Asia, the Middle East, and Latin America. China will also invest in mines in Africa, Latin America, and Asia to supply its rapidly growing industrial demand for iron, copper, and other minerals. China's own reserves of oil have diminished to such a point that it attempted to cut deals with many countries, including Iran, to gain access to their petroleum reserves.[25] Of course, China is investing internally as well. As it approached the magic year 2000, China's State Development Bank invested in energy projects fueled by hydro-, nuclear-, and oil-powered energy production plants.[26]

International investment has fostered coalescence to such an extent that the health of one nation's stock market and currencies can influence the internal stability of all other nations. The 1997 Asian currency crisis, which started as a rather innocent currency devaluation in Thailand, launched a crisis that spread to Malaysia, Japan, and other Asian countries. Many feel that the actions of top Western international currency speculators, themselves players in the international financial system, exacerbated an already unstable economic state of affairs. By 1998, European and American manufacturing companies, unable to sell products to Asian consumers, began laying off workers and middle managers.

Although this crisis abated as the world approached the millennium, it made economists, corporate managers, and ordinary citizens acutely aware of just how interdependent the electronically connected world of high finance has made us all.

The International Race for the Innovation Edge

One sign of the rapidly expanding scope of the world's production system is the tendency of previously low-tech developing countries to produce an ever-greater number of the world's inventions and technological breakthroughs. According to a recent article in the *Economist*, although spending on research and development (R&D) is still heavily concentrated in the richer economies, many Asian nations are pumping funds into R&D projects. By the late 1990s, South Korea and Taiwan were already spending as much on R&D as a proportion of their GDPs as many European nations were. In addition, these smaller countries' R&D spending is going up more quickly than that of America and Britain.

Such increases in R&D spending helps enhance species coalescence. According to a recent study at the London Business School, the creation of knowledge spreads its benefits across borders. Increased innovation in Asia should raise living standards in rich countries by expanding both the range of goods that consumers can buy and the inputs that firms can use. In addition, as innovation helps to lift Asian incomes, the market for rich world exports will expand.[27]

For years, non-western countries such as Taiwan were perceived as imitators, not innovators. Taiwanese companies, once suspected of copying product ideas and then producing knockoffs instead of inventing their own products, in 1996 obtained 2400 patents in the United States, the seventh highest total in the world! That small island now registers more patents in America each year (in proportion to its population) than Britain does. Taiwan's innovations in computer technology have enabled the country to now produce one-third of the world's notebook computers and one-quarter of its desktops. Taiwan is the world's largest maker of monitors, keyboards, motherboards, and image scanners. In addition, such innovation produces international economic synergy. This newly innovative Taiwan now is Intel's largest Asian customer for its highly prized central processing units, the brain of the computer.[28]

India is establishing a reputation as an innovator in high technology fields as diverse as software and satellite technology. India's highly skilled and well-educated workforce bolsters its reputation as a world-class producer of software. Bangalore in southern India has become a regional high-tech powerhouse, containing hundreds of high-tech companies, multinationals like IBM, Intel, and Hitachi as well as family-owned businesses. Thirty thousand software engineers reside in this city! India, which only launched its first satellite in 1975, now has a fleet of 20 satellites, with plans to launch many more. Such satellites are critical to India's goal of developing sophisticated telecommunication, television broadcasting, meteorology, and disaster-warning systems. India can forecast changing atmospheric and geological conditions with its new state-of-the-art Remote Sensing Satellite (IRS-IC). Such a development will certainly foster the establishment of species coalescence—more so because India is sharing the data generated by IRS with the United States, Germany, and Thailand.[29]

Advances in transportation and communication technology are accelerating the production and consumption synergy. Hence, not only does progress in transportation and communication technologies in and of itself bring people together; these enhancements in turn facilitate the growth of a universal production/consumption system that further connects the citizens of the world as producers and consumers.

The Planetary Communications System

The combined power of the Internet, wireless technologies such as cell phones, and other such communications are forging the growth of a planetary communications network that will link all members of the species.

These new technologies are advancing and solidifying species coalescence in some fascinating and unexpected ways. They are empowering humans to understand how strangers think, feel, and act and helping them discover the values, tastes, opinions, and prejudices of people living halfway around the globe. In a greater sense, moreover, these new technologies are nurturing a development of a holistic *gestalt*, the ability of all of us to "sense" one another, as fellow beings, members of the species. These new inventions have the potential of instilling in all of us a feeling of solidarity with our fellow humans and a sense of identification and common membership in a global community.

Historically, our ability to communicate has grown via a series of distinctive "information revolutions." The development of language itself was the first of these revolutions, enabling humans to share ideas and plan for the future. Another profound revolution, which evolved nearly 6000 years ago, was the invention of writing. The Sumerians assembled a commercial empire because they could record transactions, calculate profit and loss, and in general chronicle the development of their global commercial enterprise. Writing helped Rome to build its central city-state, organize its empire, and operate a mighty military system. The next information revolution, Gutenberg's fifteenth-century invention of the printing press, or more specifically moveable type, provided the species a method for quickly replicating and transmitting ideas, plans, sketches, maps, and other important documents. The invention of moveable type allowed Luther and other key figures of the Protestant Reformation to quickly transmit their ideas concerning God, man, and salvation to an extremely vast audience and overwhelm attempts at censorship and suppression. The fifteenth- and sixteenth-century European exploration of the Americas owes much to the new technology. In massdistributed machine-printed books, articles, and pamphlets, explorers publicized their discoveries in the Americas, generating popular enthusiasm for seeking out the New World for adventure and profit.

The Electronics Revolution and the Advent of Instantaneous Information

Inventions based on the electronic transmission of information changed human relationships even more profoundly than did writing and printing. With electronically based communications technologies we can instantaneously transmit great amounts of high-quality information over long distances. These new technologies' ability to connect intimately and instantaneously one human with another brings members of the species together in a way former information media could not. In short, electronic communications is a major catalyst in the development of species coalescence.

The telegraph, perfected by American Samuel Morse, ushered in this modern age of instant communication. For the first time, we could transmit news from New York to Missouri or Atlanta in real time, using Morse's famous dot-and-dash code. By the 1890s, telegraph wires and cables were everywhere, over land, across continents, and under oceans. The telephone, patented in 1876 and 1877 by Alexander Graham Bell, slowly led to the development of national and international telephone systems. The telephone experience was an infinitely more human interaction than the telegraph's. You could hear the voice of the person at the other end, and detect her or his emotions—displeasure, reluctance, and interest—that written words in and of themselves might not express.[30]

Radio and television ushered in the age of electronically based mass media. By the 1930s, commercial radio emerged as the first medium that could transmit messages and soon live programming to large audiences. The immediacy of radio gave it the power to bring communities and whole nations together with an intensity inconceivable before its introduction. Live cross-Atlantic radio reports of the war in London, replete with the live sounds of bombing and the blitz, helped pave the way for America's entrance into World War II. The three most influential political leaders of the 1930s and 1940s, Franklin Roosevelt, Winston Churchill, and Adolph Hitler, blessed with resonant radio voices, could use their tremendous radio "presence" to mobilize sentiments of the millions. Television, combining voice and picture, has an even greater potential for advancing species coalescence. Highly sophisticated satellite technology enables the televising of events to a worldwide audience. About 40 percent of the world's population experiences the American Super Bowl simultaneously; even more watch the Winter and Summer Olympics. Our global population also can experience in real time historic events, such as the 1969 Moon Walk, the assassination of John Kennedy, and the fall of the Berlin Wall.

Over the period of a century, these communications systems, which become ever more realistic and immediate, are changing the way we view each other and ourselves. We gradually become more like each other, or more to the point, we become less like strangers. Our feeling of empathy, or commonality, is a phenomenological experience—it does not mean that we have overcome all conflicts, or even that we necessarily like each other more. Rather, these technologies are making us less alien to each other.

By the late 1990s, we were combining the computer, advanced telephony, and satellite technology to deliver another communications revolution. The species finally had the ability to communicate instantaneously, not only by voice, but through text, picture, and video. Innovations of the 1980s such as the facsimile, or the fax, enabled people to transmit over the phone lines a copy of a document, a report, a bill, or invoice. Businesses suddenly could increase productivity exponentially. A vendor, for instance, could send a prospective client a design immediately by fax; the business could critique the design, redo it if necessary, and send it back to the vendor. Businesses now saved countless numbers of days that would have been lost waiting for mail delivery of the document. The fax technology enabled workers to complete projects much more quickly than before.

Through the last two decades of the second millennium, a plethora of new communication marvels accelerated human interaction and served as a catalyst to species coalescence. Improvements in wireless technology made it easier for people to communicate with anyone from anywhere, without depending on a phone connected to a wire running from a wall. It has become commonplace to see people talking on phones while walking, driving, and sitting on beaches. Of course, with enhanced communication came diminished privacy—the fact that I can be reached by phone while driving my car lessens the extent to which my car is my private domain. We can all now be contacted, anywhere, anytime.

Electronics is enabling us to work, and play, more efficiently. By the 1980s, businesses and universities were using satellite communications to "video-conference." Corporate executives sitting in a room in Singapore and branch managers in a room in New York could now both talk to and see each other. Corporations now could enhance their mentor programs with videoconferencing—managers in one part of the country were suddenly empowered to become mentors to junior employees thousands of miles away because they could now see their proteges without having to travel to the junior's location. Universities quickly used videoconferencing to broaden their audience and increase their enrollments through a new process called distance learning. A university professor in Montana could lecture students sitting in a room hundreds of miles away watching his image on a large screen.

The Internet, Cyberspace, and Species Coalescence

These new technologies enabled us to communicate more quickly, more elaborately, and with more people simultaneously. Nevertheless, many people feel that the communications breakthrough that will have the most profound impact on species coalescence will be the Internet and technologies derived from this new human networking phenomenon. Originally established as a system to link university and other researchers to each other so they could more easily collaborate on projects and share findings, the Internet has blossomed into a worldwide communication modality used by millions of people.

The Internet, based on computers linked through telephone or cable systems, permits ordinary individuals to tap into other computers to receive information about a plethora of subjects. According to a survey conducted by the Pew Research Center for the People and the Press, over 74 million Americans use the Internet, an amazing figure considering that in 1995 the vast majority of the public had never even heard the term *Internet*. These people use the Web for business, entertainment, and utilitarian purposes such as weather reports, travel, and vacation information. They shop online and buy stocks through online brokerage services.[31]

A new revolution, called "broadband," will exponentially increase the speed at which Internet content will be sent and received. Fiber-optic and standard cable lines will speed into homes voice, data, and video, making the Internet somewhat more like the television experience.

This is a truly international phenomenon. Europeans are just as avid Internet users as Americans are. Even developing countries are becoming interested in this new technology. The number of Internet users in China surged to 1.5 million in 1998 from 600,000 in 1997. In China the urbanites tend to have greater access to technology—China has 540,000 computers linked to the Internet and 14,000 Internet service providers, of which 3550 were based in Beijing—and 84 percent of China's Internet users are under 35 years old. China is trying to increase the number of databases and Web sites—the government is spending billions to upgrade its information resources and develop networking software. U.S. Internet guru Nicholas Negroponte predicted the number of Internet users in China would balloon to 10 million by 2000.[32]

In contrast, Islamic countries such as Iran have relatively low Internet use. Most of these nations' citizens are too poor to own computers. The country does sport a few "Internet cafes," but those are used mostly by journalists and government officials. In addition, the government itself does not seem too anxious to subsidize computer use or Internet usage among its citizens currently. Islamic governments suspect that easy access to the Internet would expose its citizens to Western ideas and images about politics, mores, and culture.[33] In 1999, a top Iranian cleric, Ayatollah Ahmad Jannati, suggested a technological alternative to censorship to combat Western electronic cultural encroachment: set up Internet sites that would transmit Islamic culture, ideas, and images throughout the Islamic world and even into Western countries.[34]

New technologies are enabling Internet users to "speak" through computers so that their voices can be heard by other users in a chat group. In addition, by placing a tiny camera on top of one's computer one can transmit video images over the Internet to any number of people equipped to receive these images. Such breakthroughs are prompting some people to run their own online "television shows." Thousands of people are running a variety of entertaining if at times bizarre home-based programs, with names such as "Vickicam" or "Allancam"—they simply place a camera in their living room and/or bedroom, and allow World Wide Web browsers—voyeurs to peer into their lives, sometimes on a 24-hour-a-day basis. More extraordinary is that some of these cam Websites receive a thousand or more hits per week.[35]

The Internet and its affiliated technologies now endow each of us with the heightened ability to contribute our creativity to the enrichment of the species. In the last few years, the technology for producing and distributing everything from text to music to pictures has become amazingly democratic. Surf the Net for a few hours and you will witness what seems like the entire world contributing information about health, money, education, science, and culture in chat rooms and newsgroups. People construct Web pages that feature their writing, poetry, art, and homegrown original music.

The World Wide Web will empower all of us to reach multitudes of people in ways and at speeds that were unthinkable even in the early 1990s. Soon, new technologies will make electronic communication even more realistic. Virtual reality will enable us to enjoy electronically the tactile and visceral experiences we usually only can have in the real world. This new form of media/communication experience is unlike any that has come before it. Virtual reality, or VR, requires that the user wear a mask or goggles that project images directly onto the eye. The user looking through these goggles sees lifelike images that appear to the user to be part of his field of vision. The goggles are attached to computers that ensure that the images the observer experiences are totally real. As the wearer moves his or her head, the "virtual" landscape, which is entirely computer-created, moves across his or her vista. Of course, the more sophisticated the program and more powerful the computer, the more authentic the experience seems to the user.

New technologies are making the VR simulated world ever-more realistic. Now we can equip the user with computer-sensitive gloves to let the person move and manipulate objects in cyberspace. Argonne National Laboratory houses a virtual-reality installation called "The Cave" where participants use VR glasses to experience almost anything they want: a nuclear explosion or a walk through a modern home.[36] Although visitors use The Cave primarily as an entertainment medium, Argonne performs serious research there on the effects and impact of the VR experience. Recently at Disney World, I took a "virtual tour" of the Vatican. In this world, my fellow cybernauts and I walked along the Vatican's corridors and through the grounds; we also visited parts of the Vatican inaccessible to the average tourist, such as the bell towers and its walls. We even had the opportunity to "crawl" along the Sistine chapel ceiling. NASA uses virtual reality technology to train pilots using advanced VR flight simulators. Moreover, virtual reality is increasingly being used as a method to train doctors and surgeons—the physician trainee operates on the virtual patient with the use of specially programmed gloves.[37] By 2000, Montefiore Medical Center was using virtual reality techniques to train prospective surgeons to perform high-risk sinus operations near the optic nerve and brain. The training experience is so lifelike that students can actually feel a scalpel cutting through the delicate interior anatomy of a virtual nose.[38]

Now, imagine sitting in front of your computer in 2005, equipped with the latest VR technology. You are wearing a mask or helmet, and gloves, all plugged into the computer. You have just logged on to a meeting held in a "room" in cyberspace. We now "sit" in this "cyberroom," speak in real time, see the other people, full bodied; we now not only converse, but can interpret their body language, grimaces, and smiles, and hear their laughter (or snickers). We will be able to speak over each other, and of course pause and engage in polite side conversations, just as we would do in person. Of course, this is all happening inside one's virtual reality helmet, clothing, gloves, and so forth. It looks, and feels, about as real as it possibly can, given the fact that we are not physically present in a room with these people.

By 2005, virtual reality may all but eradicate most of the differences between electronic communication and the in-person interface. We can only imagine how this type of experience will accelerate species coalescence. Every evening one can make new friends from different countries and different backgrounds. Add to this the possibility that regardless of whether each person in the cyberroom speaks our language, we will still be able to understand him or her. Even now, new technologies and software that automatically translate voice transmissions are breaking language barriers. With such a device another person's language can be translated for us in real time, as he or she speaks. Thus, an American who only speaks English can, with the aid of an automatic translator, communicate easily with foreigners in their own language. And Chinese and French friends will use the same mechanism to answer back in English without having to learn a word of that alien tongue. We could easily apply such software to the virtual reality conference in our cyberroom.

The Internet, wireless cellular phones, faxes, and electronic mail depend on complex land and air connections in order to operate. They also require a viable space infrastructure of rocketry and satellites to beam our signals around the globe. At present, over 800 communications-related satellites are in the sky. Several satellite constellations are on the horizon, including Globalstar (48 satellites), Teledesic (288 satellites), and Sky Bridge (80 satellites). Within a decade, there will be over 2000 of these birds at various altitudes, some as high as 1500 miles above the earth. (The usual flying altitude for a commercial airplane is about 5–6 miles.) They will be reading water meters from space, offering advanced two-way paging, tracking fishing vessels on the high seas, and enhancing Internet and videoconferencing services. These flying transmitters will enable travelers who have taken wrong turns in their autos to use the satellite-based Global Positioning System (GPS) to determine their exact location and how to reach their destination.[39]

As we approached 2000, more countries joined the satellite-launching endeavor. China successfully launched its Long March 4-B satellite as part of a joint project with Brazil to conduct environmental research.[40] Around the same time, a Ukrainian-Russian rocket carrying a U.S. DirecTV satellite was launched from an ocean-going platform situated near the equator.[41] Moreover, almost monthly, Europe's Arianespace successfully launched numerous communications satellites into orbit.[42] Soon, it will become infinitely cheaper to launch such vehicles. Kistler Aerospace, headquartered in Kirkland, Washington, has developed a rocket that, after it places its satellite into orbit, returns to Earth using parachutes and airbags, ready for relaunch within days. Such a reusable space transport vehicle will allow Kistler Aerospace, along with Kelly Space & Technology, Pioneer Rocketplane, and Rotary Rocket to revolutionize rocketry and reduce the cost of a typical satellite launch to one-tenth of its current cost.[43]

Species Coalescence and Vitalization

New communications and transportation technologies, as well as the global economy, are bringing the members of our species closer together. This trend toward species coalescence is underwriting a general global prosperity and improved quality of life. More important, though, it serves as a force enabling us to achieve what I consider a much broader goal, *vitalization*, the extension of our physical presence and human consciousness throughout the cosmos. In a host of ways coalescence becomes a means to this end.

First, on the purely functional level, to vitalize the cosmos, the many groups composing our global society must contribute their diverse skills and talents to this global enterprise. This endeavor is ultimately too immense to be accomplished by only a small segment of the species carrying humankind's banner. Certainly, over the eons, independent nations, even corporations, will embark on their own initiatives and exploratory journeys. However, to successfully pursue the vitalization endeavor, including the space mission, the entire human race must be empowered to contribute skills and creativity in a cooperative and synergist spirit. Certainly, the development of such entities as the global production/consumption system enables all members of the entire human species to significantly participate in the larger enterprise.

Second, only a fully united humankind, one that has reached a strong degree of species coalescence, can rally around a common goal such as vitalization, as well as develop a consensus over the shape and direction of the mission to achieve this goal. If our species is to invest time, energy, and resources into efforts that will lead to vitalization, such as massive space exploration and colonization programs, all parties on the local, national, and global level must "buy into" this vision.

This concept of a united humankind implies that humans will have achieved a sense of a common identity with and feel a basic allegiance and loyalty toward their species as a whole. We may rightly ask how we will achieve this sense of common identity, considering that traditionally humans have primarily directed their loyalty to, and considered themselves members of, much smaller social units—family, town, work organization, and nation.

As we can see, the technologies and processes described throughout this chapter will help us identify with and feel membership in the larger human community. Humans have a natural tendency to identify with a group, act in its behalf, and protect it when necessary. I identify with people who share my values, speak my language, and travel through the same physical space that I do, and I eventually come to share their world view and perspectives. This tendency of humans to join and support reference groups will never change. Now, however, people are expanding their reference groups to include people from widely varying backgrounds located all over the globe. This is not an unexpected phenomenon: Transportation and communication systems enable individuals to experience—touch, see, and hear—a much wider range of people than they could even a few decades ago.

The growth of international and multinational companies is producing employees who regardless of their country of origin or cultural, ethnic, and religious differences, enjoy a sense of identification with each other. This new solidarity is created as people work together in a functionally independent organization for the same goals—teamwork, as it were. This process is intensified by the fact that the political units to which many of these employees belong are themselves melding into larger entities. For instance, Western European powers such as Germany, France, and England, once bitter rivals, are now politically enjoined in a European Union with a single currency. It would be unthinkable for these countries to guard shared borders or point missiles at each other. The United States, at odds with various European countries throughout its history, is gradually becoming organizationally connected with the European continent through trade associations and such military pacts as NATO. The aforementioned new Silk Road connecting

China and Europe will, as it facilitates trade between and general communication among citizens of distant regions, eventually lead to a further global bonding.

Third, the sense of commonality and membership in the species that we are developing in this era will serve us well throughout a future in which we will be traveling over vast distances throughout the galaxy. Although our descendents might build thriving colonies and develop vibrant civilizations in the farthest regions of the cosmos, living far from Earth they might eventually lose physical contact with the rest of humanity. Regardless of how large a single distant colony becomes, it might be so far from another human settlement that it is feasible for its members to have contact only with other members of the colony. Our colonies will become our new homes, our fellow settlers our only friends. The sense of comradeship and identity our species is forging in the twenty-first century will serve as a source of coherence and cohesiveness over long stretches of time of isolation from others of our species.

Last, these relatively isolated new societies on humanity's far frontiers will be the standard-bearers of humanness—it is they who will vitalize the universe. The very concept of "humanness" that they have so strongly and permanently assimilated into their consciousness because of coalescence will serve as their guideline, and their vision, as they vitalize the universe.

Species Coalescence versus the "Global Brain" Phenomenon

In the *expansionary* view, the purpose of species coalescence is to further progress of both humankind and the universe. Advancements in transportation and communications enable greater human contact, which should then lead to greater understanding and identification. Coalescence also enables us to work together in an interfunctional manner, opening the door to a global synergy never previously experienced. The more we communicate our ideas and share our creativity, the more we can achieve as a species. The resulting prosperity and expanded human ability will empower us to vitalize the universe.

Central to the expansionary view is the tenet that individual genius and talent form the lifeblood of species' progress. Each person's creativity plays a crucial role in our evolutionary advancement. It stands to reason, then, that to contribute, each person must be free to act as an independent agent. If my individuality is stifled, my creative light is extinguished. Any attempts to suppress the free flow of information or control individuals' creative activity in effect will undermine our mission, the achievement of vitalization. Therefore, even as we move toward greater interdependence and global intimacy, as it were, we must ensure that individuality is protected.

A society that experiences coalescence is one in which the species' members gather together into a working, operating group, converging synergistically as would a well-tuned organization or team. However, this process is a coalescence, not a "coagulation." At different points in its evolution, members and subgroups of our species may even split off and pursue their own goals and at other points come back together again. In fact, the final product of the individual working alone, getting in touch with his or her own "genius," is probably as important to the growth of the species as anything a team might accomplish. The underlying principle is clear: The species' progress depends on its individual members pursuing their individual purposive self-advancement.

Over the recent decades, a contrary vision of the coalescing of the species has arisen, one that while largely unfamiliar to the public enjoys a phalanx of avid adherents. This vision, a hybrid perspective born of New Age philosophy and physical cosmology, stands as the diametric opposite of the perspective espoused in this book. The expansionary vision sees the species evolving into a group of unique individuals who share a common identity and mutual goals. The emerging New Age perspective literally looks to a day when individuals are ultimately absorbed into the whole, spiritually, mentally, and in some of the more extreme scenarios even physically.

A recent article on an Internet site, entitled the *Principia Cybernetica Web*, summed up this perspective succinctly. The article claimed that society could be viewed as a multicellular organism, with individuals, human beings, that is, playing the role of the organism's cells. The network of communication channels connecting individuals, such as the Internet, plays the role of a nervous system for this superorganism, a "global brain," as it were. The article goes on to compare the functioning of society with that of organs, systems, and circuits in the body. For instance, the industrial plants extract energy from raw materials as the digestive system processes food into nutrients. Roads, railways, and waterways transport these products from one part of the system to another one, as arteries and veins send nutrients throughout the human body. This model compares our armies and navies to a "giant immune system" protecting society against invaders.

Such analogies have over time led certain observers to conclude that individual humans may be similar to cells of a "social superorganism."* This article, which represents the core beliefs of a group of writers, thinkers, and scientists that calls itself the "Global Brain Group," contends that trends in electronic communications such as the ones described in this chapter are leading us toward "global integration." To quote: "As technological and social systems develop into a more closely knit tissue of interactions, transcending the old boundaries between countries and cultures, the social superorganism seems to turn from a metaphor into a reality."[44] The article's author mirrors the sentiments of the New Age "aquarian" philosophy when he equates international computer networks like the Internet with a "global brain" or "super-brain," the World Wide Web "with capabilities far surpassing those of individual people." He speculates that connections like the Internet will ultimately lead to "the integration of the whole of humanity, producing a human 'super-being.' "[45]

Futurist Barbara Marx Hubbard, head of the *Foundation of Conscious Evolution*, is a leading proponent of the global-brain perspective. In an interview that appeared in the *Futurist* several years ago, Hubbard asserted that "humanity is becoming part of a larger whole, both through our consciousness and through our electronic connections." The next step in human evolution is not the expansion of individual abilities but a growth of a "collective capacity, both technologically and spiritually."[46]

Hubbard expresses appealing sentiments about how each of us can expand our "human potential," attain "spiritual growth," and through human action acquire almost "godlike" abilities. However, listen closely and you hear Hubbard and other New Agers, many of whom now populate the worlds of academic physics and cosmology and influential policy groups and think tanks, betray their belief that they are not really interested in the progress of humanity, and especially the individual human. The goal of our frenetic developmental activity is not the evolution of humankind, but rather the progression of an abstractly defined "nature." Hubbard opines that "Nature evolves through the formation of larger whole-systems—from atom to molecule to cell to humans, and now through planet Earth to one interacting system. As the whole-system matures, its parts exercise synergy and become greater than they were when separated."[47]

---

* This view is not unlike the Gaia concept developed by Havelock Ellis, which conceives of Earth as a unitary, living organism. Humanity, in this schema, is a cluster of "living cells," part of this living organism. We will speak of this theory more in later chapters.

When asked what our evolution is moving toward, where humanity is actually headed, her answers become curiouser and curiouser. She says that we are becoming a "universal species" with higher consciousness. "We are moving toward a whole-centered, God-centered, universal consciousness. . . ." Again, Hubbard echoes the global-brain and universal-mind philosophy.

Many of her quasireligious statements reflect the thoughts of someone she identifies as one of her spiritual guides, Teilhard de Chardin, a Jesuit anthropologist who lived, wrote, and performed scientific research in the first half of the twentieth century. Teilhard's originality and creative approach to subjects such as evolution led him to ask penetrating and original questions about the role of humanity in the universe. Unfortunately, many of his answers, especially those related to the concept of species coalescence, are counterproductive to the human enterprise. His fundamental theories provided the framework for many who espouse a belief that humanity is evolving into superorganisms, global brains, superbrains, and related concepts.

Teilhard likened the stages of the Earth's evolution over time to the layers of its zonal composition. He would thus refer often to the metallic inner core of the Earth surrounded by rock, the lithosphere, which in turn is covered by the fluid layers of the hydrosphere (water, especially the oceans), all of which is surrounded by the atmosphere. For Teilhard the most interesting layer around the Earth is the living membrane composed of fauna and flora, and biological entities such as birds, mammals, and, of course, the human species. This biosphere is the layer of life.

Most observers have no problem with a taxonomy of the Earth based on its various layers. It is this last layer, the noosphere, which has become a source of some controversy. This is the concept that most attracts New Agers and various global-brain adherents to Teilhard's world view. Although commonly associated with Teilhard, the concept of the noosphere was actually introduced in the early 1920s by geologist Vladimir Vernadsky. Teilhard, like Vernadsky, described the noosphere as the realm composed of conscious thought: words, ideas, culture—all of those amorphous entities related to our cognitive and thinking processes. This thinking layer has emerged and evolved since the Tertiary period (about 70 million years ago) and has spread over and above the worlds of plants and animals.

The expansionary vision presented in this book totally subscribes to Vernadsky's concept of the noosphere—the realm of ideas, art, and human consciousness. It reflects an understanding that the thinking aspects of humanity are so potent that they should be taken as seriously as we do air, rock, fauna, and flora. Vernadsky conceived of the noosphere in much the same way we view the idea of culture—though we can't see culture we treat it as real, describing its characteristics and qualities and ascertaining its effects. We can easily envision the human species extending this metaphoric noosphere into the heavens as we establish human civilization on new planets.

However, Teilhard takes the concept of the noosphere to the very limits of credibility. For Teilhard, it is as physically real as the hydrosphere and atmosphere. He proclaimed that this thinking layer, is truly substantial, so much so that it emits its own form of energy. This *radial* energy is a product of the thoughts and ideational activity of the noosphere. Such energy would be different from that with which we are familiar, the *tangential* energy that serves as fuel for power plants and automobiles. Teilhard posits that this radial, or psychic, energy will eventually come to dominate the planet and in the far future become independent of tangential energy. In other words, he seems to suggest that our thinking selves will detach from our physical selves. This noosphere, the sum total of the human thought process, would gradually evolve into a supersapient being, which he labels the "Omega Point." Individuals as we commonly understand them would be merged into this supersapient being.

A close reading of Teilhard's books, such as *The Phenomenon of Man* and *Vision of the Future*, reveals that while Teilhard seems to be describing the progression of the human species, his real focus is on the development of what he labels "life." Teilhard makes clear that the human species' primary role is to serve as an agent for evolution of this "life." Humankind is only a stage in this movement of life toward its ultimate end, the "supersapient being," that is, the Omega Point. For Teilhard, this Omega Point is actually "the end and the fulfillment of the spirit of the Earth."[48]

Of course, those supporting a Teilhardian vision of the maturing noosphere are much excited by the most recent developments in communication technologies that have developed in the last 30–40 years. The fact that the Earth is now ringed by satellites, millions are communicating hour after hour on the Internet, and television and radio continuously broadcast information has convinced many that the noosphere is not only thriving but also already approaching a critical mass. The age of the global brain must be right around the corner!

Other technologies on the horizon will only heighten their expectations that the formation of the noosphere is reaching such an apex. Recently British Telecom announced that it would pour untold millions of pounds into the development of a new technology called the "Soul Catcher." Supposedly,

British scientists are developing a concept for a computer chip that, when implanted into the skull behind the eye, will be able to record a person's every lifetime thought and sensation from birth on.[49] Dr. Chris Winter, head of an eight-scientist artificial-life team at BT's Martlesham Heath Laboratories near Ipswich, predicts that within 30 years the technology will exist that will enable people to relive other people's lives by playing back a record of that person's experiences. Winter goes further—he claims that we could re-create a person physically, emotionally, and spiritually, at least as a computer program. Winter said "Soul Catcher 2025" would enhance communications beyond current concepts. "I could even play back the smells, sounds and sights of my holidays to friends." At a 1995 World Future Society conference lecture in Boston, Dr. Ian Pearson, a colleague of Winter at British Telecom, mentioned that such "plug and play" technologies, which enable us to communicate our thoughts through the Internet, would be equivalent to an electronic form of telepathy.

One can see how news of this possible development could very well light a fire under the global-brain contingent. After all, this Soul Catcher technology will enable us to instantaneously share our every thought, maybe our lifetime of thoughts, with everyone else plugged into the universal network we call the Internet. Moreover, we will have the ability to receive the impulses transmitting others' lifetime of experiences stored on their Soul Catchers at the same time they are receiving our stored impulses. Eventually, the sum total of everyone's emotions, impressions, and images transmitted through neural communication with and over the supercomputing neural Net would coagulate into a global brain. We would come exceedingly close to achieving Teilhard's Omega Point.

This vision of humanity melding into a supersapient being is as opposed to the expansionary vision as is humanly possible, pun intended. From the expansionary point of view, we work to achieve species coalescence in order to better coordinate our productive activities, and at the same time nurture in all individuals a strong awareness of who we are and what our destiny is. We will then be prepared to actualize our destiny to spread consciousness and humanity throughout the universe.

The underlying assumption of *The Future Factor* is that without the unique contributions of self-developed, purposively self-advanced human individuals, there will be no further progress of the universe. If as the new-age types contend individual humans ultimately become absorbed into a larger supersapient entity, individuals will lose their power to create beauty and innovate. No individual creativity, no progress, on Earth or anywhere else! A recent article in *The New York Times* on the Soul Catcher spoke of the potential negative impact of such systems on individual creativity. The writer warned that as our electronic networks evolve into "a sort of global nervous system," the plugged-in brains will find it difficult to "sort, sift, and prioritize the constant deluge of information." He worried whether our brains, pummeled by the onslaught of data and information, will "degenerate into data land-fills."[50]

However, from the New-Age/Teilhardian viewpoint, what does it matter? After all, it is the creation of the superentity that counts, not the quality of input and output. "Data landfills" may be exactly what Teilhard thought individuals might have to become in order for this supersapient Omega Point to emerge. You only worry about the quality of the content of individual minds if you feel that human creativity and invention have any role to play in the evolution and perfection of the universe. If the ultimate goal of the coagulation of human minds is as Teilhard says, the evolution of the "tree of life" and the "fulfillment of the spirit of the Earth," the fact that individuals become data landfills is a secondary, if not altogether superficial, concern.

This New Age superrational global brain is more than vaguely reminiscent of the *Borg* on the later *Star Trek* series, a superrational entity that cruises the universe absorbing other life forms. Of course, while this disturbingly robotic entity claims that it absorbs life forms to "improve" them, make them more rational, we instinctively know that this entity operates on the basis of what C. S. Lewis would call a "hideous strength." It squeezes the life out of all life forms, and turns them into superrational, or should I say totally mindless, lifeless robots. Global-brain enthusiasts reject this analogy. In 1999, Peter Jennings, on ABC *World News Tonight*, hearing British Telecom's Dr. Peter Cochrane describe Soul Catcher, asked "are we on our way to becoming 'Borg'? Are we in danger of being 'assimilated by the collective'?" Cochrane responded that "as a species we are in general far too rebellious, irreverent, and individualistic to become the Borg." Considering how Soul Catcher would actually operate, I find it rather disingenuous for Cochrane to so cavalierly deny the possibility that such technologies could end individuality as we know it.[51]

At least the Borg, grotesque and austere as it is, has a physical manifestation. The Omega Pointers never tell us that Teilhard saw our destiny as "detaching the mind, fulfilled at last, from its material matrix, so that it will henceforth rest with all its weight on God-Omega."[52] This new being would exist, but not physically! The only "image" I can possibly muster to imagine what Teilhard saw as he wrote these words is *cyberspace*.

What possibly puts the Teilhard/New-Age vision most at odds with the expansionary vision is nestled deep in the mechanics of how the noosphere must evolve into this supersapient whole. According to Teilhard, in order to reach this Omega Point, this layer of thought must achieve "compression"— it must achieve a physical critical mass. Compression can only occur in a limited physical space. Therefore, to reach the Omega Point, humanity must be kept living on a sphere (i.e., the planet Earth) in "the geometric limitation of a star closed, like a gigantic molecule upon itself."[53]

Teilhard's scientific analysis is clear: Humanity cannot evolve into this supersapient being and simultaneously move out into the cosmos. If humanity expands into the cosmos, we can never achieve the critical mass of thought Teilhard states as necessary for us to merge into a supersapient being. Humanity, as it travels through space, over millions of miles if not light years, will be projecting itself in all directions. We will communicate with each other over the light years, but will be too dispersed, to ever achieve this compression of thought, this noospherical critical mass, frustrating any attempt to complete the "circle" Teilhard so desires. That is why Teilhard stated that space travel would hinder what he considered the future evolution of humanity.

The expansionary vision claims as the destiny of man vitalization, the extension of humankind's consciousness and intelligence throughout the cosmos. In Chapter 6 I describe at length the profound linkage between the evolution of both humankind and the cosmos. If humanity ever evolved in accordance with Teilhard's vision, into a supersapient Omega Point, it would never achieve its wondrous and extraordinary destiny to vitalize the cosmos. The consequences for humanity, and the universe, would be devastating.

Why the Expansionary View Will Triumph

We find ourselves in the midst of this battle between perspectives—as the world turned 2000 the *Futurist* on its front cover proclaimed the emergence of "the Global Brain." The modern media seems obsessed with stories of the melding of mind and machine.

However, there are several reasons that this "group mind" ideology will not triumph. For one, too many individuals of the Expansionary Vanguard, as I label the legions of pro-human and pro-growth people, are actively campaigning for interplanetary exploration. This in and of itself will get the species off its duff and prevent it from spending eternity contemplating its cybernetic navel.

Sheer economic necessity will also trigger a movement off planet Earth. The "thought" and knowledge industries, Internet-based companies included, cannot in and of themselves support human civilization. Our species' energy and "hard" manufacturing-based industries must continue to thrive if we are to progress. We will eventually turn to the spheres—planets and asteroids—for our energy resources and other material rewards. In short, we must turn to space travel and exploration if we are to improve our standard of living on Earth. Without an ever-growing array of Earth-orbiting satellites, we could not hope to maintain our modern defense, banking, or telecommunications systems. Space is a place for humanity to live and prosper—it will beckon us long before we become absorbed into the noosphere's supersapient being.

In the final analysis, the expansionary world view will overwhelm the seductive global brain concept because the human species has always and will continue to strive to meet its true destiny, vitalization. We all sense that we are made for something better; we have the ability to transform the Earth and the universe, bringing light to darkness, transforming the cosmos into a thing of harmony and beauty. This sense will drive us to bring life, including elements of Earth's natural environment, to dead or moribund spheres, and spread human consciousness throughout the cosmos.

Species coalescence is one of the processes by which we will get ever closer to reaching this destiny. In that sense, we pursue species coalescence for the same reason we direct our efforts in such activities as biogenesis, cybergenesis, and dominionization. We will never allow species coalescence to drift into the formation of an amorphous global brain.

Now, let us begin to explore our destiny.

5

Vitalizing the Universe

In the Heavens as It Is on Earth

Throughout the book, we have explored how our mastering of forces such as biogenesis and dominionization is preparing us to achieve a much greater goal, vitalization. We will now take a closer look at this process and survey humankind's efforts to achieve this magnificent goal on Earth and elsewhere. Later I will present the case for considering vitalization not just a goal, but actually humanity's destiny.

I use the term *vitalization* in its most comprehensive sense. According to the dictionary, to vitalize is to endow something with spirit and vigor, to animate that object, energize and activate it. By vitalization I mean all that, and more—we will strengthen and rejuvenate the galaxies and arouse them to life by implanting life, or rather the *life force*, throughout the universe. Certainly, the term implies establishing in the universe the *physical* presence of human beings. Moreover, we will introduce *humanness* to the cosmos—we will bring order to the heavens, as well as the human concepts of progress and growth.

While vitalization is something that we will perform "out there," throughout the universe, let us remember that we have been engaging in this process on Earth for most of our history. We have brought order, consciousness, and all that is uniquely human, to our planet. Though Earth did spawn humanity, humanity has now become the major player in the future evolution of the planet's ecosphere. We have cultivated nature, nurtured the growth of other species, and harnessed and transformed many of the core physical processes.

For the last few decades, we have been vitalizing areas beyond the Earth, especially the ionosphere, that region a hundred or more miles above the Earth. We have transported our intelligence into the heavens, first with communications satellites and soon with such permanent structures as the International Space Station. We are rearranging the sky, making it a place where humanity can work and live.

Our entry into the world of the outer bodies such as planets and asteroids, as explorers and colonizers, will be a key event in humanity's effort to improve the universe. For the first time human consciousness will begin to insinuate itself in the universe and spread organic life onto what have been barren worlds. We will extend the human presence to the Moon and Mars, to the solar system; and over the centuries and eons throughout the entire universe itself.

Humankind is involved in a veritable maelstrom of activities aimed at vitalizing the universe—the International Space Station, the Mir space mission, and the new and relatively unknown Astrobiology Program from NASA, to name just a few. Both government and private interests, such as corporations and the Mars Society, are planning to make Mars the first major target for true vitalization. In addition, we are now using an intriguing new concept called *terraformation* to replicate on Mars the climate and topography of Earth.

A Torrent of Activity

As we entered 2000, China's Xinhua news agency unexpectedly announced that China had launched its first shuttle into space. It launched the spacecraft Shenzhous (*God Ship* or *Divine Vessel*) from a base in northwest China on a more powerful version of its celebrated Long March rocket, which it had been secretly developing since 1992. China had also been constructing a super-modern space control network, which would guide and monitor the spacecraft as it circled the globe 14 times before touching down in the northern province of Inner Mongolia. China now joined the United States and Russia as the only nations in history to launch a vehicle capable of carrying a man into space.[1]

China followed this launch with another spectacular announcement. According to Zhao Bing of the China Academy of Launch Vehicle Technology, "The day the first Chinese goes into space is just around the corner." Some in the Chinese scientific community predict that China will launch a human into space by the end of 2000, while Western observers put such an event at least 2 to 4 years into the future. In any case, it is certain that China will become the third country in history to send a human spacefarer into the cosmos.[2]

Such events indicate that the drive to achieve vitalization is rapidly gathering steam. Think of the rash of space-related projects that have been undertaken in only the last few years that have been added to an already illustrious list of space enterprise. The space shuttles (especially one launch featuring 77-year-old national hero John Glenn), an interplanetary exploration mission known as the Cassini project, and the continuing adventure that is the International Space Station have focused the public's attention on space and its potential benefits to humankind. Space telescopes deliver new astronomical findings so frequently that we have come to expect our vision of the nature and evolution of the cosmos to undergo radical change without prior notice.

The Cassini represents one of the most ambitious space probes in human history. Launched in late 1997 by the United States, Cassini will be exploring the solar system through 2008. Cassini will travel past Venus twice and finally go into orbit around Saturn. By July 2004, Cassini is scheduled to send a probe called Huygens onto the surface of Titan, one of Saturn's moons. The Huygens probe will gather information about Titan's atmosphere and look for signs of microbial and other forms of life. Close-up photos of the icy surfaces of many of Saturn's moons are expected to be more detailed and certainly clearer than pictures from earlier missions. We will also learn much about the chemical composition of Saturn's rings. By late 1999, Cassini was snapping crisp photos of the Moon, on schedule for its 2004 rendezvous with Saturn.

We cannot underestimate the impact this mission will have on human consciousness. We are extending a part of ourselves, an intelligent machine, to the surface of distant spheres. More importantly, Cassini's cameras will be graphically revealing a segment of the universe previously hidden from humankind. As Huygens descends it will be snapping pictures of the moon's surface, which is covered with liquid methane and ethane. Humankind will get a breathtaking view of one of the only shorelines in the entire solar system besides those on Earth![3] Although Cassini carries no human beings, our species is still well represented on the mission. Mounted on the exterior of Cassini is a disk covered by a thin aluminum patch to protect against the blows of micrometeorites. The disk contains a wealth of Earth memorabilia: signatures of 600,000 people from 81 countries, postcards, letters, drawings, music, baby footprints, paw prints, and toddlers' scrawls. Technicians had scanned the signatures and other Earth memorabilia onto the digital disk before it was mounted on the aircraft.[4]

We will explore the universe through other missions. Deep Space 1, a small but smart, self-navigating probe, will be propelled all the way to an asteroid by an ion engine at speed of more than 62,000 mph.[5] Another probe, the Wide-Field Infrared Explorer, better known as Wire, was carried into space aboard a Pegasus-XL rocket in early 1999. This NASA satellite carries a telescope designed to shed light on the history of star formation.[6]

In early 1996, Japan announced its long-range vision for space exploration, which included constructing a moon base and establishing a Mars colony.[7] Within 2 years, Japan began to make good on its intentions. On July 4, 1998, to help commemorate the successful landing of the U.S. unmanned craft Pathfinder on Mars, Japan launched its own craft to Mars. The half-ton probe, named Nozomi (hope), was boosted on Japan's most powerful launch vehicle, a newly developed M-5 four-stage rocket. Nozomi is currently scheduled to begin beaming back photographs and experimental data from the Red Planet 2003.[8] The mission's discoveries will prove invaluable. Carrying 14 different kinds of instruments from five countries—Japan, Canada, Sweden, German, and the United States—Nozomi will unlock mysteries about the Martian ionosphere and upper and lower atmosphere and help us understand the impact that solar winds have on the planet. This probe may tell us whether there the planet's surface contains water, a question we must have answered before we send humans to live on the planet.[9] Japan will share data generated by the probe with scientists from around the world.

Vitalization-oriented projects such as the Nozomi mission will have an enormous impact on species coalescence. As nations work and plan together to achieve vitalization, we identify ourselves not only as Chinese or Americans but also as earthlings involved in exploring and changing the cosmos. This spirit of international cooperation manifested itself at a recent meeting of the International Astronomical Union, the official organization of the world's leading astronomers. Participants revealed official plans to turn the Moon into a nesting site for humans. Scientists with the European Space Agency, as well as those from Japan, disclosed plans for a series of ambitious space missions to establish unmanned radio telescope listening bases on the Moon in the near future. Such telescopes would operate free of the man-made "noise" that plagues Earth-based radio telescopes. Norio Kaifu of Japan's National Astronomical Observatory revealed that 10 or 20 years from now humans will be sent to the moon to help operate and maintain these telescopes. The funding for such projects will come from the Japanese government, the European Union, as well as industry, private citizens, and possibly public investors.[10]

Such projects will never succeed without public support, financial and otherwise. NASA and other government agencies are developing a host of new programs to directly involve the public in the vitalization mission. For example, America's brand new Mars Millennium Project is challenging over 100 cooperating organizations to "imagine a community yet to be created for 100 earthlings in 2030."[11] Organizers of this project chose Mars as the destination for our "100 earthlings" mainly because the public has shown considerable interest in missions to the planet such as the Mars Pathfinder mission and Mars Surveyor mission.

The Mars Millennium Project has now become a national arts, sciences, and technology educational initiative underwritten by a wide variety of institutions, including the White House, the U.S. Department of Education, NASA, the National Endowment for the Arts, and the J. Paul Getty Trust. The Clinton Administration even weighed in on the issue—in 1999, Hillary Clinton and officials of the National Endowment for the Arts announced plans to have school children, along with scientists and technicians, help design a permanent Mars settlement for the year 2030 through the Mars Millennium Project. This kind of project creatively engages the public to start seriously thinking about the future, or more precisely the human future, in a serious way.[12] Companies such as Lucent Technologies linked their scholarship development programs to the Mars Millennium Project in order to engender in budding scientists and technicians an enthusiasm for space and space colonization.

The next generation will experience space in a whole new way, as planners, travelers, and eventually inhabitants. Let us look at a few examples.

The International Space Station

Certain to accelerate the vitalization effort is the construction and operation of the International Space Station (ISS). One of the most daunting endeavors in human history, when completed the ISS will enable scientists to conduct experiments and probe our universe in ways that would have been unimaginable just a few decades ago.

Sixteen countries are cooperating to build the nearly $20-billion space station, which will eventually be the size of two 747 jetliners. The station will be completed in the next few years. In fact, the first crew is expected to begin living on the ISS sometime in late 2000.[13] The project will pave the way for later missions to the planets and beyond. "The Space Station," said Arnauld Nicogossian of the National Aeronautics and Space Administration, "is basically a test-bed to conduct future exploration."[14]

The future of manned missions within the solar system will depend on how well people fare in seemingly inhospitable environments. Our experience on the ISS will teach us how people adapt to living in space and how international crews work together. We will learn the complexities of building large structures in space replete with life support systems and medical research facilities.

This project demonstrates how quickly a megaproject can be completed when the money and the will are applied to it. According to NASA director Dan Goldin, by 2004 U.S. station development efforts will near completion.

In many ways, the level of cooperation exhibited by so many countries in planning and implementing the ISS project is almost as impressive as the technology itself. Participants are willing to act more like citizens of the world than of their own nations in order to make this project a success. Russia and the United States are not the only players in this game. Canada is contributing a Canadian Space Station Remote Manipulator System, or "Robotic Arm." The European Space Agency (ESA) plans to provide critical crew rescue vehicle components. Italy, Japan, and the Brazilian Space Agency are all contributing manpower and hardware.

While NASA and other countries' space agencies see ISS primarily as a center for research, living, and experimentation, they also hope the space station can become a moneymaker. To quote Dan Goldin: "The ISS represents a platform in space of unprecedented capability. We envision that it will become a seed for emerging commercial activity in the coming decade and we are moving ahead to ensure this outcome."[15] His goal is to have the ISS serve as a marketplace foundation—he believes that the station will stimulate a national economy for space products and service in low-Earth orbit.

The ISS will build on the success of the Mir space station project, an endeavor many considered the crowning achievement of the Russian space program. Over the course of its 14-plus years in orbit, Mir has generated valuable information. We learned how humans could live in space for protracted periods without suffering serious physical or emotional consequences. The Mir experience taught the species how to carry out complicated repairs in space, replete with the shuttling of new crews, spare parts, and even computers, to the space station on almost momentary notice. The mission showed us how to efficiently handle seemingly devastating onboard crises, such as loss of power and computer-system failure.[16] It also demonstrated the viability of tapping the international skill base to ensure mission success—America, France, Germany, as well as Russia, contributed crew members to the Mir mission. The crew that served on Mir in 1999 consisted of two Russians and a Frenchman.[17]

In early 2000, just when Mir was supposed to be decommissioned, a Netherlands-based financial group, Mircorp, unexpectedly pumped $20 million into the station in return for the right to use Mir for commercial ventures. Mircorp, which expects the space station to remain operational for several more years, has stated that it will invest up to $200 million in the station. Mircorp will use Mir as a space portal to deliver Internet data content, including pictures of Earth and the solar system. Mircorp has enlisted the help of the William Morris Agency to identify mass media and entertainment merchandising opportunities for Mir. Eventually, the company may convert Mir into a "space hotel" for wealthy tourists, a subject we will have more to say about in the next section.[18]

Mircorp wasted no time in reinvigorating the Mir space endeavor. On April 6, 2000, two Russian cosmonauts arrived on Mir to begin a 40-plus day mission partially funded by Mircorp. By September 2000 a second Mircorp-sponsored mission will bring the hardware to set up the Internet space portal.

The ISS project is enabling us to vitalize the regions around the Earth—the ionosphere, the stratosphere—in a way never before possible. We are in effect extending the noospheric activity, the thought, the creative work, the intelligence, the very consciousness of humanity into a permanent position miles above Earth. I emphasize the "permanence" aspect because, while Mir has performed admirably over the years and will continue do so for the next several years, the International Space Station represents a completely novel experience for the human species. On the ISS, humanity will be establishing a new center for creation and invention, a new locale for human activity. Moreover, because it is permanent, it becomes our species' first real home off the Earth's surface. We will not just work there—we will live on this sphere. Babies may even be born on the ISS!

Visionary Gerard O'Neill once suggested that the human species construct "space islands" at the farthest reaches of space and the solar system, man-made asteroid-sized enclosed spheres on which thousands would live. Such space islands would bridge the gap, fill the spaces, as it were, that exists between planets and within galaxies. Decades from now the ISS may well be remembered as the breakthrough prototype for such man-made miniplanets and miniasteroids.

The Advent of Space Tourism

The Frommer publishing house is known for its many travel guides. For decades they have helped tourists navigate their way across Europe, Asia, and the Americas. As we entered 2000, the Frommer company added one more tour book to their illustrious collection: *Frommer's The Moon: A Guide for First-Time Visitors*. Frommer's newest *Guide* provides prospective visitors to the moon details on the cost of going to the moon and the best places to acquire training for the trip. The *Guide* lets us know how to eat and sleep during the trip and includes a crater-by-crater guide to the most famous lunar attractions. Comments by the astronauts give us an insider's view of life on the moon.[19]

The people at Frommer are anticipating the growth of a new industry, space tourism, and take this industry's potential seriously enough to provide a tourist's guide to a sphere other than Earth. Frommer, it turns out, is correct in its assumption. The millions, perhaps billions, of people who will be populating the universe will do so not only to mine and colonize distant spheres, but also to relax and be entertained.

We know that the desire and inclination to travel to space bubbles beneath the surface of our cultural waters. One survey found that about 60 percent of Americans and 70 percent of Japanese would like to visit space. With the Hubble telescope mission, Mir, and various Mars' missions generating enthusiasm in the public for space travel, space tourism's time may have arrived.[20]

One vision of this industry features "places in space," usually orbiting the Earth, visited by people for stays of one or two weeks. They would go there for a variety of reasons, such as fun, enhanced knowledge, a great view. One can easily imagine such vacations becoming the ultimate status symbol— anyone "touring" space would be one of the few humans to have ventured off this planet.

The first site for space tourism will probably be a space station such as the ISS, which could make a few modules available for purposes of tourism. If people show an interest in going to space for vacations, entrepreneurs will surely build leisure-oriented structures in the stratosphere. One proposed facility would house 24 tourists. This craft would come equipped with living quarters, bathrooms, showers, computers, and other amenities. It might also feature an auditorium-sized room large enough so that its inhabitants could play various team sports like basketball, soccer, and whatever else they might invent for a low-gravity environment. Initially, such excursions would have a high tab, about $300,000 per person for a two-week stay. Over a 15–20 year horizon, prices could fall by 75 percent. Some think that retirees might choose to live there full-time if these environments could accommodate year-round living.[21] Japanese entrepreneurs are now openly talking about building a hotel in space. The media is already anticipating this new industry—the 1998 movie *The Fifth Element* featured a space resort circling the Earth that featured gambling, entertainment, sporting events, and a spectacular view of our wondrous planet. My guess is if we build it, they will come.

Space tourism may start off with jaunts simpler than these extravagant two-week stays in space. Several companies are signing up potential tourists to spend a few hours in space. Part of these companies' enthusiasm is sparked by a $10 million prize offered by the X Prize Foundation. The first company that can build a vehicle that can bring civilian passengers 62 miles up to the edge of space and back down again safely, then perform the stunt again within two weeks, will win this money. (The X Prize is really aimed at prodding someone to invent the reusable rocket.) Many people are signing up with the companies trying to develop the space planes to perform the trick. According to the X Prize Foundation's chief operating officer, MaryAnn Gregg, "All the research shows there is a large, latent market of people willing to spend a significant portion of their salaries for a space flight."[22]

One company offering a quick but breathtaking space experience is the Civilian Astronauts Corps. Its vehicle will take off from the Gulf of Mexico, blast literally out of the water and head on a trajectory 70 miles above the earth. (A commercial jetliner cruises at around 5 to 7 miles above the earth.) Those on board will experience 4 minutes of zero gravity, and will feel a force of 5 Gs (five times the gravity of Earth's) for approximately 30 seconds on descent. Of course, they will remember the view of Earth for the rest of their lives. The Civilian Astronauts Corps has signed up several dozen people at $3500 per trip. Another company, Zegrahm Space Voyages of Seattle, began selling spots for a week-long space adventure/education program that will culminate with a suborbital flight on a six-passenger "space cruiser." The cost for the classes plus flight is $98,000, and 30 people have already deposited the $5000 necessary to hold a spot for the flight's scheduled 2001 liftoff. Most of these companies are recruiting passengers with almost no publicity or advertising.[23]

Can we include space tourism within a concept as magisterial as vitalization? Most decidedly, yes. When we travel to space to enjoy and relax, we are vitalizing the cosmos as much as when we set up new governments and new mining and production facilities on these planets and asteroids. We are humanizing the universe, if we recognize that the concept "human" includes nuances within the human experience such as fun, diversion, leisure, creativity, playfulness, inventiveness, and other not so obviously pragmatic sentiments and activities.

And various forms of space tourism may be the method to best engage the nonscientist public in the exploration and colonization process. Space tourism, for all its frivolity and hyperbole, will become an important link in the vitalization chain.

NASA's Astrobiology Program

America's space agency has suddenly and unexpectedly created a new program that should hasten our march toward vitalization. In 1998, NASA announced it was initiating a program labeled the Astrobiology Program, dedicated to the study of life in the universe. While its focus does not directly reflect all aspects of the expansionary vision presented in this book, sufficient congruity exists between the two to warrant special mention here.

The program's main purpose is to study the origin, evolution, distribution, and destiny of life in the universe. The founding document defines life in the broadest sense, not just human existence, but microbes, flora, fauna, and other such entities.

In trying to unravel some of these mysteries, the NASA Astrobiology Program has developed a number of "canonical questions" that it feels must be answered, in a multidisciplinary fashion, to discover life's place in the universe. One question asks, "How do habitable worlds form and how do they evolve?" Devolved from this question are a number of subissues: What, then, is a habitable world, how did it get to be that way, how long has it been habitable, and how long will it remain that way? We know that Earth has been hospitable to some form of single-celled life for some 3.8 billion years. Moreover, our own solar system has several candidates as habitable worlds, including Mars, and the moons of Jupiter, Europa, and Callisto. However, these spheres do not exhibit the full breadth of phenomena conducive to sophisticated life forms that we find on Earth. NASA will dedicate itself to looking for signs of life throughout the solar system.

NASA also wants to determine how we can recognize biospheres, how living systems emerge, and the potential for survival and biological evolution of living beings beyond their "planet of origin." Another related question involves the role that rapid changes in the environment play in the evolution of a planet's ecosystem.

Two issues NASA addresses seem particularly germane to the focus of our quest. One is the influence that Earth and its biosphere have had on each other through the centuries. Astrobiology Program researchers believe that while Earth indeed spawned life, life itself has also had an incredible impact on Earth and its environment. Even plant life on Earth has had a pivotal role in the sequestering of carbon dioxide from the atmosphere. As we shall see, to formulate a fully elaborated expansionary vision of human development we must understand just what impact human presence has had on the development of Earth's biosphere. Such information will then help us determine what impact humans will have on the ecosystems of the planets and asteroids we will occupy in the future.

NASA's empirical research on another of its core questions, "Do living planets seed nonliving planets?" should clarify another aspect related to the expansionary vision. Our theory examines the contention that material from distant planets and star systems that bombarded Earth's surface eons ago might have played a role in the establishment of life on Earth. NASA's conclusions should help many researchers and cosmologists take a fresh look at this controversial concept.[24]

In any event, NASA is beginning to seriously explore issues relevant to the larger question of human destiny while pursuing its more immediate goals of establishing communications systems and monitoring the weather.

In recent congressional testimony, NASA director Daniel Goldin clearly explained how we might research a number of these questions about the nature of life in the universe. This entirety of the Astrobiology Program is subsumed under the Origins Program NASA initiated in the 1990s. In the third millennium, NASA will vigorously pursue the Origins mandate. New missions will include the launching of powerful telescopes in an effort to find the earliest structures in the universe and look for new planets. Robotic probes sent to Mars, Europa, and other targets would search for the beginnings of life in our own solar system. To quote Goldin, the goal is "to do what no generation before us has been able to—understand our place in the cosmos."[25] Origins in its totality will support initiatives to search for planets around other stars, study galaxies and stars as they are born, and to look for evidence of life elsewhere in the solar system and the universe.

The NASA program will help us answer many questions. My only concern is that the program is not proactive enough. Yes, the program hopes to "discover our destiny." However, the philosophy of those administering and implementing this program may bias the focus of our research. For instance, there seems to be too much chatter about how we can use the Astrobiology Program to discover "intelligent life" on other planets or in other galaxies. While some may find such research salient, even crucial, it tends to undermine our own sense of destiny. (I will have more to say on our obsessive search for aliens in Chapter 8.) In addition, many observers, including those championing the Astrobiology Program in space organizations such as the National Space Society, seem overly concerned about discovering the "symbiotic" relationship between Earth, the universe, and mankind. "Where does humankind fit in the great chain of being?" they seem to be asking. However, the emerging vision of human destiny detailed in this book sees the human species from a completely different context. We do not hope to discover our "place" in the universe, but rather seek to create that place for ourselves. In the human future, we the species determine the shape, evolution, and future of the universe.

However, these are minor criticisms. The Astrobiology Program is a giant leap forward for the space community in terms of its research focus, underlying philosophy, and commitment to the big questions.

Endeavors such as the ISS and the Astrobiology Program signal that our species is highly motivated to expend time, energy, and resources in an effort to extend ourselves outside of the Earth's environment. Yes, the space program has its functional rewards. Many of these planets, moons, and asteroids promise to be rich in energy and minerals, which will be returned to Earth as the prizes of exploration. Yet through our efforts we are also expressing a need, a propensity, a compulsion perhaps, to extend our presence, to populate the universe with life, to beautify, to embellish, to bestow our consciousness on the spheres.

More interesting is the fact that these efforts are only a small part of the movement just now gathering momentum to accelerate the vitalization process.

Mars: The First Vital Step

Humankind will initially vitalize the various regions of the solar system. Within that solar system, the planet Mars may be the first site the species explores, colonizes, and attempts to achieve its most ambitious feat, *terrafor-*

mation. Mars will be the test case for whether humankind at its current level of technological development can actually vitalize another sphere.

Current timetables developed by NASA, private scientific organizations, and others in the space industry have mankind setting foot on Mars by the year 2014 or so. NASA and other nations' space agencies have established this as a serious deadline. We mentioned that other countries are interested in pursuing such a project. Japan has embarked upon a nascent Mars space mission, and Russia has had an interest in the red planet for decades.

Why Mars?

Mars possesses a sufficient number of qualities to make it the leading candidate for humankind's first serious attempts to colonize and live on another sphere. First of all, Mars is close enough to Earth that we can travel there, using current rocket engine technology, in about 6 months' time. Second, it has a terrain and natural environment more suitable for habitation than do other nearby planets. Third, scientists believe that Mars may contain minerals and other elements that will enable our settlers to live off the land. During the months they spend on the planet, humans could use the elements on Mars to manufacture such necessities as fuel as they explore and chart the Martian landscape.

In addition, Mars has sparked interest among anthropologists and paleobiologists—they suspect that the Mars soil may tell us about how a planet evolves, or in the case of Mars, did not develop. Some feel that Mars should have become what Earth is, a repository of lifelike entities, even microbes. Mars did not develop life, and scientists would like to know why.

One of the attractions that Mars holds for many nations is the possibility that it may contain minerals useful in industrial production back on Earth. One government official had made the claim several years ago that if Mars were as close to the Earth as the moon is, nations would already be staking territorial claims on the red planet. Because of its complex geological history, many speculate that Mars could contain vast reserves of silver, gold, palladium, iridium, platinum, and a host of other metals useful back on Earth. Our analysis of the composition of meteorites tells us that asteroids that exist in the so-called Main Belt between Mars and Jupiter contain tons of strategic metals, such as cobalt and platinum, to name just a few. Mars, then, would be attractive not only for its own resources, it could also serve as a convenient launching site for exploratory, mining, and colonization missions to these asteroids.

There is one commercial resource that is known to exist in profusion on Mars—deuterium, the heavy isotope of hydrogen. Deuterium is the key element in first- and second-generation fusion reactors and is a staple of the current nuclear power industry. Its rarity on Earth pushes its value to over $10,000 per kilogram. Once commercial fusion reactors come on line, this element's price will increase even more. Obviously, thriving Mars colonies will be able to mine this element, ship most of it back to Earth, and use the rest in their own Martian fusion reactors. This makes a Mars colony a win-win situation, one in which Earth receives a precious resource, and the Martians become economically self-sufficient.[26]

A century from now countries that do not embark on serious Mars missions may suffer the same fate Portugal did when it abandoned its ambitious navigation and exploration expeditions in the late fifteenth century. In the mid-fifteenth century, Portugal was the premier exploring nation, the first country to circle the Cape of Good Hope around the southern tip of Africa. Henry the Navigator became world famous, and Portugal grew rich through trade with nations of the East. Yet it never pursued a serious program of exploring and colonizing the Americas. Meanwhile, Spain and then England, Holland, and France became rich and powerful through earnest exploration and eventual colonization of the New World.

If we are to colonize Mars, we must take into account this planet's topography and climate. Mars is about 4200 miles in diameter, half the size of Earth. It is 142 million miles from the sun and about 50 million miles from Earth. The planet is covered by dry, reddish-brown desertlike regions, with caps at its North and South Poles that may contain frozen water. Mars has the solar system's tallest mountain and the deepest canyon. The Martian year is 687 days long and, like Earth, features four seasons. Our astronauts will see a pink sky during the day, a blue one at sunrise and sunset.

The atmosphere, though containing traces of oxygen, consists mostly of carbon dioxide. Our human colonists and explorers will not be able to breathe such air. In addition, Mars is extremely cold—while its day time temperatures have been known to get as warm as 63° F, at night the temperature can drop to as low as $-130°$ F. Mars' turbulent winds reach several hundred mph, whipping up dust storms that last for weeks. And astronauts spending extended periods in low gravity can encounter numerous problems—a 200-pound Earthling becomes a 78-pound weakling when living on the surface of Mars.

The Profile of the Mars Mission

How will our species establish itself on Mars within the next 15 years? Well, the positive news is that this journey will be a lot cheaper than we thought only a decade ago. Our estimates have shrunk from the hundreds of billions of dollars a few years ago to now about $20 billion, or about $2 billion per year for 10 years, about what NASA plans to spend on the International Space Station. Moreover, the most recent feasibility studies from NASA's Johnson Space Center in Houston suggest that some of the proposed timelines are too conservative and pessimistic. JSC engineers feel they might deliver astronauts to Mars within 9 or 10 years.

The Mars mission as currently conceived will serve as a shining example of smart planning and common sense. The Jet Propulsion Laboratory in Pasadena, California, will send probes to the Mars surface, such as the 1997 Pathfinder and new launches in 2001 and 2003, in order to determine what technologies it will need to send humans to the planet Mars a few years later.

In 1989, U.S. President George Bush attempted to jumpstart the Mars space mission. He made a dramatic speech committing the United States to landing a human on Mars by 2020. In response to that speech, enthusiastic NASA engineers quickly devised an elaborate plan that would get us on Mars in 30 years. The space community, however, was quickly disappointed. Within the year, the Bush administration became distracted by the war in the Persian Gulf, the recession, and political campaigns. The Mars mission languished.

The plan NASA devised in response to Bush's speech did little to help the mission's cause. The resulting action plan for the Mars mission typified NASA's tendency to think big and expensive. The $450 billion price tag turned off government and the public. In addition, the plan was absurdly complicated, sporting orbiting construction hangars and space docks, massive lunar bases and transportation fleets, and oversized interplanetary spaceships.

Since that time, NASA has learned that the best way to maximize the chances for success of a mission as ambitious as putting humans on Mars is to follow simple and cost-effective strategies. After all, the purpose is to get the human species to Mars, to vitalize the planet, not to satisfy the fantasies of techies and Trekkies. In the mid-1990s, the Johnson Space Center's Exploration Office came up with a plan called the Mars Reference Mission that could get humans on Mars for about $50 billion. Then, in 1997, they announced that they could cut the mission price tag in half, mainly by eliminating the need to design a brand new launch vehicle. NASA's Mars Reference Mission brings down costs by using existing technology whenever possible and reducing the tonnage of all the material we will send into space, including the space craft, fuel, payload, and the like.

Kent Justin, chief engineer in JSC's Exploration Office, states that the underlying operating principle of the NASA Reference Mission approach is to first determine the month and year that that they want the crew to arrive on Mars, and build a strategy by working backward from that point. Using such a logistical framework, the planners have determined that they must start preparations for the mission about 26 months before the astronauts actually set foot on Mars. NASA determines what the mission will need in terms of rockets and cargo ships, to ensure mission success. According to Joosten, "We send out as much of the critical materiel as we can, and make sure it gets there, make sure it's operating, before we send the crew."[27] Earthbound engineers will transport necessary materials to Mars and robotically set up camp before the crew ever gets to Mars.

We also have to perform our launches while Earth and Mars are relatively close to each other because Earth and Mars circle the sun at different rates. They align in relation to the sun and each other only about every 26 months; we enjoy the luxury of this advantageous alignment for only a few weeks. If we launch our vehicles during this 2–3 week window of opportunity, the craft can travel from Earth to Mars in as little as 6 to 10 months. Once they have landed on Mars, our explorers will have to wait on Mars about 15–20 months for the next window to open that will enable them to return home in a similarly short time.

Here is one scenario for getting our people on Mars quickly, efficiently, and safely. In 2011, we launch two rockets to send payloads to Mars. One payload is the mission's Earth-return vehicle, the craft on which the astronauts will fly home. It will stay in orbit around Mars until our astronauts are ready to use it for their homeward voyage. The other rocket contains all material the astronauts need to live on Mars: the surface habitat and its power systems, a fuel-production system, rovers and other exploration equipment, and the Mars ascent vehicle. Amazingly, once this vehicle has landed, it's going to be producing propellant for the ascent vehicle, fuel for the rovers, and air and water, mostly out of the Martian atmosphere.

These frenetic activities will continue for a full two years before the crew ever lands. During that period, the habitat will be robotically set up on the surface. This living abode may be inflatable, a recent innovation designed to reduce the payload weight. Finally, and before the astronauts are sent on their way to Mars, mission control will confirm the readiness of the entire

Martian base for human habitation. In short, we will do all that is humanly possible to ensure that the space base infrastructure is working properly. We reduce risk to as near to zero as can be expected in a mission of this magnitude and complexity.

According to NASA's grand design, in 2014 a crew of six will be launched from Earth on a mission that will last about 2½ years. They will spend six months getting to Mars, about 500 days on the surface, and finish the mission on a six-month voyage back to Earth.

The sheer length of these early astral expeditions will make these trips serve as a test, not just for the astronaut but for all humanity. The Mir space mission has already provided us the opportunity to investigate the many physical challenges involved in living in space for protracted periods. However, in a Mars mission we are traversing whole new landscapes in the social and psychological realm. What are the mental and physiological effects of long-term flight, of three years or more, facing unknown dangers, in a totally alien physical environment, no less? Moreover, what are the psychological impacts of residing on a planet tens of millions of miles from the home planet Earth. These spacefarers will be pioneers in more ways than one. The success or failure of these first Mars navigators will have momentous implications for the human species and its ability to meet its destiny.[28]

Some might ask why we do not just send robots and unmanned vehicles to explore, photograph, and possibly mine the red planet. In 1997, the unmanned Pathfinder, with the help of the 22-pound Mars rover called Sojourner, photographed the Martian landscape and sent back valuable data about Mars' rocks and soil. From this data scientists have been able to deduce that the Martian interior has experience a long history of melting, cooling, and remelting.

However, robots are limited in what they can accomplish on distant spheres like Mars. Even with advances in computer miniaturization, scientists can squeeze just only so much artificial intelligence into a probe as small as Pathfinder. Therefore, these robots will have to be commanded from Earth. However, the great distance between Earth and Mars makes it impossible to instantaneously command such a robot. As we entered 2000, we also learned the mechanical limits of such robotic probes. The Mars Polar Lander, which was supposed to land on Mars' surface and search for the existence of water under the planet's surface, suddenly stopped communicating with Earth as it descended to the red planet. For several days, the Earth-bound technicians helplessly sent out signals to the Polar Lander before they gave it up for lost or destroyed.

Most scientists familiar with the vagaries of space exploration are certain true exploration of Mars will require humans who can roam over the surface of Mars and do what the human brain, not a computer, can do best, peruse leads and hunches in a way that machines cannot. Their view would seem to contradict Moravec's contention that "universal robots" could ever have the capability to explore the cosmos, make independent decisions, and even plan their next missions. In a recent article in *Ad Astra*, the official organ of the National Space Society, two NASA scientists make the point that robots are most suited to perform activities preprogrammed by humans. Greg Schmidt and Michael Hawes write that with humans on Mars in a central base camp, surrounded by highly capable robots, "we could both extend our exploration to the entire planet as well as dramatically shorten our search for evidence of life." Humans on Mars would utilize robotic reconnaissance airplanes and land rovers to perform this exploration.[29]

On these missions, humans will do what no robot can do: exercise their inherently human trait to decide what on Mars is interesting and beautiful. While on Mars, our astronauts will be charting the surface of Mars, creating a cartographic representation no telescope on Earth could draw. They will be growing food, using one of the habitats as a chamber for cultivating a diverse array of plants. They will perform experiments in low gravity, and produce food. Mostly, though, they will be learning, and teaching the rest of us, how to actually live on a world other than our own.

This mission, after all, is not a one-shot deal. It will provide the infrastructure for a permanent base on Mars.

Even before our crew departs from Mars, crafts carrying equipment for a second human mission will have reached Mars. These crafts, in orbit around the Red Planet, will carry such items as a second Earth-return vehicle and a second surface habitat. This floating equipment, intended for use by the next crew, for the time being will serve as backup equipment for the first crew in case their equipment on the ground malfunctions. At the end of their 500-day sojourn on the Mars surface, the spacefarers will use liquid oxygen and methane produced on the surface of Mars to fuel their ascent vehicle. They then will link up with the Earth-return vehicle that was earlier put into orbit around Mars and ride that spacecraft back to Earth.

Getting the mission to Mars has its own challenges, and will probably require a new type of launch vehicle. This vehicle is called the Magnum. The process envisioned by NASA is complicated, some believe overly so. In its simplest sense, the Magnum would put the components of the Mars mission into low-Earth orbit, and the astronauts would assemble the parts while in orbit in preparation for the trip to Mars. From there they would continue on to Mars.

The scientific community is feverishly debating which type of engines our spacecraft should use to travel to Mars and other spheres. The relationship between dominionization and vitalization are quite clear here. The greater the control we have over the forces of nature, the easier it is for us to develop the energy needed to move ourselves off the planet. Some have suggested using solar-electric propulsion, enlisting the power of the sun to propel our spaceships. Others think that rockets propelled by nuclear power could make the journey to Mars in 100 days, not the six months a chemical rocket would take. And proponents such as Samuel L. Venneri, NASA's chief technologist, point out that the onboard reactor could then be used on the Martian surface to provide power to support a base on Mars. Lasers could propel our vehicles into the furthest reaches of the solar system. In addition, many feel that tapping the huge power of fusion to propel spacecraft would serve as a way to propel our spacecraft. A spacecraft with a fusion reactor on board would possess enough energy to reach any solar system destination with ease.[30]

The Emergence of the Mar's Underground

The effort to get the species to Mars depends on more than the minds and energy of university scientists and government bureaucrats. Humankind will also make the critical step in our achievement of vitalization through the efforts of private individuals and organizations devoted to making Mars colonization a reality. Some organize conferences and seminars to spark public interest in this mission; others raise money to finance space launches.

The resuscitation of the Mars project demonstrates how a small group of devotees can keep the fires burning for an idea or a project even when official policy circles have lost their enthusiasm or government leadership falters. After the Apollo mission's successes, including the successful moon landing in the early 1970s, the U.S. government's original support for a Mars flight cooled. It cut back on funding for new programs, and imaginative planning disappeared. NASA concentrated its remaining energy and resources on the space shuttle program.

However, since 1981 an eclectic group of space enthusiasts has been meeting annually in Boulder, Colorado, to keep alive the dream of sending mankind into space. In 1981, the very first "Case for Mars" conference was held, organized by a group of graduate students who had very little idea what their sessions would lead to. By summer 1998, the conference membership had grown to over 700 people. At that point, they formed a Mars Society to actively lobby for a Mars program. They even signed a Mars Declaration, which summarized their belief that human flight to Mars had at last become feasible and more desirable than ever. They sent copies of their declaration to Washington and posted it on their Web site (www.marssociety.org.) for the public to read and evaluate.

At the 1998 conference, a parade of speakers presented evidence that we could land people on Mars and establish a permanent presence there. From a ragtag organization in the early 1980s had evolved a well-organized support group whose conferences were attended by NASA scientists and astronauts, university researchers and graduate students, and even ordinary citizens from all over the world.

One of movement's leading organizers is physicist Robert Zubrin. A former senior engineer for Lockheed Martin, Zubrin now heads his own aerospace engineering firm, Pioneer Astronautics of Lakewood, Colorado. Zubrin has used his prolific writings, as well as organizations such as the Mars Society, to introduce ideas on space travel that have had a significant impact on theory and research in the field.

Zubrin has designed a Mars mission strategy called Red Planet Mars Direct, which he says will be cheaper and faster than NASA's. Zubrin's plan would send a crew of four, not six as NASA is planning, in a craft smaller and lighter than NASA's. He would begin his mission in 2005 with the launch of a vehicle similar to the Magnum. The craft would carry an Earth-return vehicle directly to the surface of Mars, without stopping to orbit Earth. A direct flight, Zubrin says, would eliminate the need for a space tug or any other additional hardware in orbit.

NASA likes some of Zubrin's ideas, but does not agree with the size of Zubrin's planned quarters for living on Mars, a much smaller abode fitting only four crewpersons. The agency has, however, been influenced by one of Zubrin's central concepts, that astronauts can live off the land once they reach the Martian surface. He believes that the Martian surface contains many resources astronauts can tap to supply their life-support systems. In addition, Zubrin believes that Mars can provide the fuel for the astronauts' return trip home. The NASA space missions in 2001 and 2003 will investigate a number of factors critical to the Mars Direct and Mars Reference success, including discovering whether Zubrin is correct in his contention that we can produce fuel on the surface of Mars.[31]

At the conference, a myriad of other speakers spoke on a wide variety of issues, from the engineering and medical challenges of a Mars mission to the legal and philosophical issues raised by human exploration of the planet. At one of the conference's sessions, a young NASA engineer, George James, promoted the idea that local resources on Mars itself could be utilized for everything from habitat construction to fuel development. Some of the sessions at the 1998 conference described new surface transportation schemes. At one meeting, a few aerospace engineers exhibited their blueprints for small unmanned airplanes capable of flying across vast expanses of the Martian terrain. These planes could survey complex geological regions, including many that look too rough for surface rovers.

Many sessions dealt with one of the key issues related to space flight, exploration, and colonization, specifically the sociological and psychological aspects of living in space for protracted periods of one to two years or more. One session described problems that might emerge as astronauts on Mars tried to communicate with those on Earth. Depending on the relative position of Mars and Earth, a radio signal sent from one orb can take anywhere from 8 to 30 minutes to reach the other. Therefore, people on each orb cannot talk to each other in real time. My message to you on Earth takes 8 minutes to reach you. Your reply also takes time. Obviously, humans must devise novel methods to speak to each other in such time-distorted conditions if we are to effectively communicate between Earth and Mars. (Imagine trying to handle a Martian-based medical emergency from Earth with an 8-minute plus delay between questions and answers.) At the conference, simulation exercises using 4-minute delays showed that humans could adapt to this new communication rhythm after an hour of practice.

Zubrin and his colleagues are using the force of will and the expansiveness of their vision to bring about historic change. The Mars Society, with the help of Zubrin's energy and zeal, is trying to raise $1 million for the construction of a Mars habitat module in the Canadian Arctic. "We want to have that up and running by 2000. If we can do that, we can raise perhaps $5 million to $10 million to fly a hitchhiker payload to Mars by 2003," Zubrin says, perhaps on a planned European Mars probe. The Society is even trying to raise enough money to fly a robotic mission of their own to Mars, without the assistance of any governmental group.[32]

As we can see, the exploration and colonization of Mars and other orbs will present challenges the species has never encountered. However, our species is planning to take an even bolder step in its efforts to vitalize the universe. From the minds of scientist and space enthusiasts alike comes the most audacious idea of all: to not only travel to Mars and establish a space base there, but to literally make the surface and atmosphere of Mars just like those of Earth itself!

Terraformation

Nothing is more emblematic of the concept of vitalization than the emerging plan to not only implant human intelligence and consciousness on another world, but to physically transform that sphere into a human-friendly environment, suitable for life and all human activities.

As fantastic as it may seem, many scientists are convinced that we can convert the environments of planets such as Mars from their current lifeless state into something much more "earthlike." On a terraformed planet, humans could live, work, and even grow food in the open air. We would not wear space suits, nor live in domed cities with artificially produced air.

The Terraformation Process:
A Myriad of Visions

There may exist thousands, or even millions, of orbs throughout the universe suitable for terraformation. In our solar system, Venus, Saturn's satellite Titan, and Jupiter's moon Europa have been suggested by scientists as the bodies that might be considered sites for terraformation.

Mars, though, contains the most inclusive list of conditions that make it suitable for terraformation with our present technology. A planet, first of all, must be somewhat similar to Earth if we are to transform the sphere. We mentioned earlier the similarities between Mars and Earth—it has a nearly 24-hour day, there is a possibility that water exists on Mars, and it has four seasons. It is so "earthlike" that we suspect that 2 billion years ago it may have contained some life forms.

The scientific case for terraforming Mars was first made about 20 years ago and has been gaining momentum ever since. In the 1970s, the Mariner and Viking missions to Mars revealed that the planet's surface, while containing no life, possessed all the chemical elements needed for life to exist. Two biologists of the National Aeronautics and Space Administration Ames Research Center, Maurice Averner and Robert D. MacElroy, were inspired by the data generated from these missions to investigate whether Mars' environment could somehow be made hospitable to Earth-based life forms. Their studies suggested that it could.

Over the years, scientists applying ever more sophisticated climate models and ecological theory have borne out Averner and MacElroy's contention that Mars' physical environment could at least theoretically be made human friendly. People have developed a variety of models for changing the red planet with widely varying timetables for when this would happen. Some of the more prominent theorists in the area are Martyn J. Fogg and the Mars Society's Robert Zubrin. In 1996, I attended a lecture Fogg delivered at a National Space Society meeting, and was impressed with data he presented validating the feasibility of terraforming Mars. His book, *Terraforming: Engineering Planetary Environments*, has become the bible of this nascent science. In his book he provides guidelines for the study of terraformation, the "ecopoiesis" of Mars, and the terraforming of Venus, and he presents a variety of standard methods and fringe concepts in terraforming a planet. Zubrin described strategies for terraforming Mars in his widely read *Case for Mars*.[33] Another proponent is Christopher P. McKay, a NASA Ames Research Center astrophysicist who has conducted many studies in this area.

In the broadest sense, the terraformation process would proceed along the following lines. The first act must include somehow warming this incredibly cold planet. Therefore, many suggest that we deliberately trigger a massive global climate change in order to raise the atmospheric pressure and surface temperature to create a "shirtsleeve" environment for humans. Terraforming would begin by introducing enough greenhouse gases, such as chlorofluorocarbons (CFCs), into the atmosphere of Mars so that trapped solar heat would raise average surface temperature by about 4°C. If that technique sounds to you suspiciously like global warming, let me assure you, you are 100 percent correct. When your planet is ice-cold, global warming is a positive development.

We would create the CFCs needed to start this warming process by setting up a plant and manufacturing them. The factory needed to produce the CFCs would require about 1000 megawatts of power, which could be produced by one large nuclear power plant. The source of the CFCs would be elemental constituents, mined right from the Martian air and soil. (Possibly another advantage for using nuclear-powered spacecraft to get to Mars.)

At this point, the slight rise in Mars' temperature would be enough to cause some of the carbon dioxide frozen in the south polar cap of Mars to vaporize. The melting caps would add carbon dioxide—itself a greenhouse gas—to the atmosphere. These additional greenhouse gasses would further raise surface temperatures, which in turn would melt more of the frozen gas. This melting would continue until the entire polar cap was vaporized.

Now we have managed to get the temperature of Mars up another 70°C, certainly within human comfort levels. The heat would also melt ice to start the planet's waters flowing once again and raise the atmospheric pressure to one-fifth of sea-level pressure on Earth, equivalent to that on some mountaintops on Earth. Humans easily survive in such environments, though on Mars they would still have to wear scubalike breathing devices to supply oxygen. However, thanks to the higher temperatures and corrected atmospheric pressure, they could dispense with bulky space suits and pressure domes.

In a recent *Scientific American* article, Christopher McKay of Ames Research points out that at this juncture, we will have already made the planet a much "kinder, gentler place than today's Mars," at least in so far as human habitation is concerned. We have not yet completed the terraformation of Mars, but at least we are halfway there. We would have enveloped Mars in a thicker carbon dioxide atmosphere, warming the planet above the freezing point of water. Add a bit of nitrogen to the atmosphere, and suddenly plant and microbial life will be able to thrive.[34]

Because plants would be growing in this atmosphere, we could cultivate farms and forests on Mars' surface, thus providing food for human colonists. In addition, the plants would slowly be "oxygenating" the atmosphere. (I would not be surprised if at that point we would begin to bioengineer the future humans to make them more adaptive to this semiterraformed environment, either through bionics or genetic manipulation. I'm sure we would be tempted to see if we could tinker with the human body to meet Mars' environment halfway.)

How long will the entire process take? Some feel that a true oxygen-rich atmosphere will have to wait hundreds of years. The plants must gradually take in the carbon dioxide and produce sufficient oxygen. McKay's computer models claim that we could generate a thick carbon dioxide atmosphere on Mars in about 100 years and create a water-rich planet in about 600 years. However, Fogg, Zubrin, a host of other scientists, as well as the terraformation boosters in organizations such as the Mars Society and the Millennial Foundation, think that new technologies may speed the initial warming process. One such suggestion has us placing huge orbital mirrors around Mars and reflecting large amounts of sunlight onto the planet surface. The quicker we warm the planet, the reasoning goes, the quicker we complete terraformation.

Of course, such timetables and predictions are made by people whose view is grounded in current technological paradigms. However, scientific breakthroughs always surprise and confound. Our description of the dominionization process is really the story of humankind continually exceeding its own definitions of the possible. Over the next few decades, current research and experimentation with the laser, HAARP's microwaves, and the mysteries of electromagnetism may generate breakthroughs that could so enhance our control over matter and nature itself that quickly transforming a planet's environment may become relatively easy.

McKay, surprisingly, interjects a bit of counterintuition into the discussion of the timeline for terraformation, suggesting that even if we could dramatically accelerate the terraformation process, perhaps we should take a slower approach. He speculates that working with longer time scales would allow life on Mars to adapt and evolve and interact with the environment—as has been the case on Earth for billions of years. Slowing the process of environmental evolution would provide us an opportunity to study the coupled biological and physical changes as they occur. This is as much a research project as it is the creation of a living habitat, says McKay—learning how biospheres are built is part of the scientific return for the investment in bringing Mars to life.[35]

Terraformation and Its Role in the Vitalization of the Universe

As I have contended throughout this work, the ultimate goal of the human species is the vitalization of the universe. This concept is multifaceted, yet in its essence it represents the expansion of the quality of humanness to all spheres. We are already engaging in this process on our home base Earth. We strive to extend human consciousness, human intelligence, and other such anthropic properties into nature itself in order to increase the intelligence and order of the universe.

Terraforming other spheres, transforming their environments into ones that can support life, certainly qualifies as a form of vitalization.

Let me state here my belief that terraformation will not be the only way for the human species to exist throughout the cosmos. It may be the preferred state of affairs, and where we can achieve an Earth-like environment, we will.

Certainly, over time Mars, Venus, and some moons within our solar system will be terraformed. However, even as we endeavor to perfect the terraformation process and apply it on planets such as Mars, we will be establishing our physical presence on spheres in a myriad of other ways. In the twenty-first century, advances in rocket technology will bring us vehicles propelled by lasers, nuclear fusion, and other methods that will hurtle the human species at lightning speed well past Mars. Long before terraformation is perfected, we will have reached distant spheres—asteroids, planets, and the like—and established human colonies, living and thriving in immense biospheres, underground mall-like cities, and a host of other structures as yet unimagined.

The point here is that all arrangements and structures that enable humans to expand their presence in the universe represent vitalization. Terraformation would certainly make it easier for the human species to live on other spheres. However, we can achieve vitalization while living on these planets in non-Earth-like environments, such as biospheres. It can even take place in totally artificial space habitations. Our own International Space Station is not a terraformed environment—it is an artificial, totally fabricated structure. Still, humans living on the ISS will be vitalizing the Earth's ionosphere by introducing rational structures and human creativity into that environment. So too would the pioneering space colonists living on the aforementioned space islands.

Therefore, under no circumstances need we wait for the species to perfect the terraformation process before projecting ourselves across the solar system and beyond and populating hosts of planets, asteroids, and moons. We should go as far as our technology at any given time can take us, and we should live in those new areas in any way we see fit. When the terraformation of Mars is complete, we should wherever feasible then apply this now-perfected technique to the many planets and moons we already populate.

Having said that, let me stress in the strongest words possible the absolute necessity of eventually perfecting this technology. We should do so even if we develop methods to live perfectly wonderful and comfortable existences in biospheres and underground cities. Terraformation involves the ultimate act of creation, bringing something that is dead, moribund, to life. Terraformation, when viewed as the infusion of being into nothingness, becomes the quintessential act of vitalization.

What do we mean by "creation," though? Certainly, terraformation is an act of scientific creation. The actions described so far—increasing carbon dioxide levels and melting polar ice caps, raising temperatures, generating heat—are essential to the success of this enterprise. However, a description of such actions does not tell me what the new land will actually look like. Even to say that it will "look like Earth" tells me little about the final product. What part of Earth? Africa? Ohio? The Baja Peninsula? Most discussions in the literature on terraformation omit asking the most important question: What kind of world will we create?

The reason we cannot answer such a question is that there is no final or right answer. Terraformation is only partially a scientific enterprise. Having mastered the scientific and technical principles of terraformation, we will discover that the process is largely an artistic endeavor.

At its core, the terraformation projects over the next centuries will represents artistic achievements of an unsurpassed magnitude—the ultimate landscape painting, the quintessential architectural project, as it were. Our species will be building mountains, creating lakes and oceans, and developing the contours of an entire new world. Such enterprises will require that we transcend the scientist in us and let the artist thrive. No two terraformed planets will look the same. Form may follow function, but only to a point. The human spirit will invent a thousand different shapes and apply a thousand different colors to these planetary canvasses!

To create is to affirm our humanness!

In the beginning, we will change these spheres into the image and likeness of Earth. Yet, as in all art, novel artistic and architectural trends will predominate, only to be replaced by the next millennia's visions.

How Vitalizing the Cosmos Will Change Mankind

As we vitalize the planets, including terraforming them, we change the cosmos. However, in a thousand ways, our efforts at vitalization also change the human species.

Of course, some of terraformation's effects on the human species will be more obvious than others. Through terraformation, our species will be able to live on other spheres with a minimum of discomfort. Living in Earth-like environments will have the same impact on human civilization that other geographic expansions and colonizations have had throughout history. Many space enthusiasts like to remind us how the European migration to the Americas changed the migrants' views of themselves and their institutions. As the colonists experienced freedom and autonomy in the New World they came to question the centuries-old legitimacy of monarchical authority. In addition, the self-reliance they developed enabled them to consider new forms of governing such as democracy and republicanism.

Likewise, we will see a plethora of new forms of institutions and social experiments arise as humans live in their terraformed environments.

Vitalization will also strengthen and reinforce the four key processes that enabled humankind to achieve vitalization in the first place. In the chart appearing at the front of the book, you will notice that while arrows point from processes such as biogenesis to vitalization, arrows also point back from vitalization to biogenesis and the other core processes. Let us look more closely at this feedback effect here.

The four processes described in the first part of the book all contribute to our ability to achieve vitalization, especially our efforts to extend our presence to Mars and other worlds. Dominionization helps us master the skills necessary to change the topography and climate of other spheres, as well as develop the capabilities to travel throughout the universe. Through biogenesis, we are constantly upgrading the physical aspects of the human being to modify the human body so it can survive in the myriad of new and challenging environments on other spheres. Through cybergenesis we will develop the cybernetic tools—computers, robots, computer implants—that will endow us with the added brainpower and computational skills we need to reach the stars. Species coalescence enables all members and groups composing global society to work together, contributing their diverse skills and qualities. Importantly, species coalescence will help humankind establish a common understanding of and agreement about the shape and direction of vitalization.

Vitalization feeds back into these processes to improve our efforts in all these areas. In the dominionization area, our foray into space will help the species develop new forms of energy. The moon and Mars contain certain elements which will allow us to quickly develop energy sources that we may never have developed if the species chose to remain on Earth. While species coalescence is a necessary prestep to successful vitalization, the very exercise of working together in concert to achieve such large projects on Earth and in space serves to bring nations and diverse peoples closer together. The activities surrounding the multinational enterprise known as the International Space Station will actually foster global cooperation. The United States and its former enemy Russia will engage in joint missions to launch Mars probes and eventually explore and colonize Mars. These efforts will soon include Japan and the European Space Agency. In turn, the ESA is proceeding with initiating a multinational effort known as the Lunar European Demonstration Approach to soft-land a spacecraft on the moon as a prelude to lunar exploration and colonization. Thus the vitalization process will help the species further transform itself from a disparate group of adversarial factions to a more cohesive entity, a force that can control its own development and its own progress.

Our efforts to vitalize the universe will also accelerate progress in the biogenesis process. Space exploration and colonization will eventually impel the species to genetically engineer the human organism into several variations of the original *Homo sapiens* model. We will be required to adapt ourselves to a wide variety of planetary conditions and reengineer the human organism to increase its chance of survival on space voyages lasting for several years. In addition, the special conditions of new space environments, such as microgravity, will provide the perfect environment for us to manufacture special classes of pharmaceutical products that will improve human health.

Vitalization's effect on these four processes is indeed profound. However, it effects another area of our development perhaps even more. I am referring to the probability that vitalization will significantly transform the human psyche. As we succeed in our efforts to vitalize the universe, which will include such momentous events as the terraformation of planets such as Mars, we will also be increasing our ability to create, invent, and innovate.

Let me explain why this will be so. The achievement process operates within fairly predictable parameters: The more confident a person is that he or she will achieve a goal, the more likely he or she is to attain it. Psychologists call this a sense of self-efficacy; some call it simply self-confidence. In addition, the process tends to reinforce itself. The more successes I enjoy, the more confidence I have in myself that I will be able to succeed in the future. Parents will occasionally let a child win at a game such as chess so the child can experience self-efficacy, knowing that once the child believes he or she can master the game, he or she plays better. Even standard business texts advise managers to permit employees to enjoy a taste of success early on to keep them focused and motivated.

In all these examples, the people involved were experiencing a progressively enhanced vision of what they could accomplish. Moreover, this does not just occur on the individual or small work-group level. It happens to whole societies as well. As our achievements as a species grow more momentous, our self-concept of what we can achieve, at the species and the individual level, becomes dramatically enhanced. Observers have always been amazed at how once people overcome the psychological barriers that stand in the way of achieving a certain goal, they routinely achieve that goal in the future.

For example, for decades professional runners tried to break the infamous 4-minute barrier to running the mile. Theories abounded asserting the physical limitations of the human being in breaking this barrier. Then, in 1951, an Englishman Roger Bannister ran a mile in slightly under 4 minutes. Suddenly, it became commonplace for Olympic runners to run at these previously unattainable speeds. Why could runners routinely break the 4-minute mile after Bannister did it? Had the human species suddenly undergone a dramatic physical change in 1951? Did human evolution take a subtle digression that year, changing the physiology of the human species? No! Once Bannister set his record, the species' concept of what human runners *could* achieve changed forever; we demonstrated the uniquely human capacity to completely reconfigure what we objectively have considered the rules of nature.

Within the last 150 years, we have exponentially increased human capacity in all areas: We live longer, travel faster, and command forces of nature that were considered the domain of the gods. And each achievement dramatically enhances what we *believe* we are capable of achieving next. Our invention of the combustion engine and our harnessing of electricity led us to believe that we could develop energy sources that could eventually break the bonds of gravity. Now, we routinely send satellites hurtling through the cosmos. The human of 200 years ago could wish to fly. Nevertheless, he could not realistically imagine the species ever would.

A subtle process is at work, which I argue is the real critical component of human endeavor. This principle can be stated quite simply: Every successful human act dramatically enlarges the species' concept of what humanity can achieve in the future. Every accomplishment induces us to seriously believe, no, inherently recognize that we can scale scientific and technological heights that seemed unreachable only a day, a week, and a year ago. Each success establishes the psychological and sociological conditions for further triumphs. At the deepest level, such accomplishments profoundly enhance our definition of the human being.

Tom Crouch, senior curator of aeronautics at the Smithsonian Institution's National Air and Space Museum, succinctly describes how the Wright brothers' first flight in 1903 changed the way humans saw themselves and their capabilities. Airplanes lifted not only our bodies—they also elevated our imaginations. "The real impact was on our vision as a human species,"

Crouch says. "Before 1903, you heard, 'If God had meant for us to fly, he would have given us wings.' After 1903, people said, 'If humans can build a machine to take us into the air, what can't we do?' "[36]

Which brings us to the issue of what impact the eventual achievement of such miracles as terraformation will have on the human concept of the possible. At its core, terraformation represents the process by which the human species infuses the dead planet, the moribund moon, with life. What was dark, we made blue and green. What was dry, we made vaporous. What was moribund, we made vital. As was the case in our previous human achievements, our first successful terraformation will change our concept of the doable. However, this time the "doable" is the ability to fundamentally change the universe itself.

The moment that humans on Mars finally take off their oxygen masks and breathe the Martian air that they have created, walk through Martian terrain that they have landscaped, and sail on the oceans they have brought forth, they will have a completely different vision of their species and their future. At that instant, the mental and sociological conditions will be created that will set the stage for an explosion of human invention and achievement that will dwarf even terraformation.

A humankind that finally comprehends what it can accomplish will know no bounds. What will we do next? Create artificial suns to heat and nourish both the man-made space islands and revive natural but dead planetary systems? Move planets across galaxies?

However, one more surprise awaits us. Our next actions will be influenced not just by what we think we *can* accomplish. They will be ultimately directed by what we know we *must* accomplish. And at that point the most powerful component of future human activity will emerge and predominate: the realization of humanity's destiny!

6

The Expansionary Vision of Human Development

Humanity is on a quest to improve the species, marshal the forces of nature, and reshape the universe. The question that still remains unanswered is, of course, why? What motivates humankind to feverishly prepare itself for what seems to be the grandest of missions? Humankind, a species residing on an infinitesimal island in a corner of the universe, dares to believe that the fate and future of the universe lie in its hands. What act of pride do we commit, what hubris do we exhibit to entertain the notion that we even have a destiny, let alone such a lofty one? Moreover, who are we to believe that we not only possess such a magnificent destiny, but also are capable of mastering the skills and knowledge necessary to fulfill such a mandate?

I present now what I label the *expansionary theory of human development*, which seeks to answer such questions. I formed this theory by synthesizing ideas currently being debated in anthropology, physics, theology, sociology, cosmology, and other fields. This synthesis borrows some ideas from new and radically unique paradigms, such as complexity theory and the anthropic principle, which are revolutionizing our views of humankind, our evolution, and our future. However, the roots of this perspective hearken back to the mid-nineteenth and early twentieth centuries.

The Emerging Sense of Destiny

Throughout this century, many thinkers, philosophers, scientists, and writers have offered visions of the future of humanity, which, taken as a whole, signal the emergence of a brand new cosmology. While these new perspectives have largely gone unnoticed by the public and the media, as we begin the third millennium, they are ready to step out of the shadows to gain recognition as a legitimate explanation of the human condition.

These novel theories share a few common elements. First, each focuses on the unique qualities and abilities of the human species, especially our extraordinary gift of consciousness. These new theories also concentrate on humanity's ability to master the environment and perfect nature. More importantly, this revolutionary new cosmology endeavors to conclusively establish the special role that humanity plays in the universe.

There are several reasons such visions have emerged throughout the twentieth century and seem to be increasing in number as we begin the twenty-first. For one, humanity has become more capable of impacting the physical world through our technology, knowledge, and science. We wonder about our *role* in the cosmos because we are quickly acquiring the ability to *change* the universe. Second, over the last century, astronomical revelations are revealing to our species just how vast our universe is. Telescopes such as the Hubble present us a picture of a teeming, endless melange of stars stretching millions of light-years. On one level, the expansiveness of the universe overpowers us, convincing us of our lonely existence. On the other hand, as this gargantuan cosmos unveils its wonders, it seems to entreat us to make it our home.

The Impact of Russian Cosmism

We tend to think of the development of the scientific approach to the art of prognostication as a recent phenomenon, an activity largely housed in think tanks, institutes, and universities. Earlier futurists, such as Jules Verne and later H. G. Wells, were essentially novelists and science fiction writers. Although they accurately predicted television, satellite communications, space travel, and world wars, we would hardly characterize their predictive methods as either rigorous or scientific.

However, in the late nineteenth and early twentieth centuries, a considerably more disciplined approach to the study of the future emerged. This field was simply labeled *cosmism*, cosmism philosophy, space philosophy, or in deference to its national origins *Russian cosmism*. Its practitioners used both theoretic inquiry and empirical research to explore the history and philosophy of the origin, evolution, and future existence of the universe and humankind. Cosmism drew from both Eastern and Western philosophic traditions. The eclectic nature of the movement's "membership"—philosophers, physical scientists, artists, religious thinkers, and poets—ensured that it would maintain a rich and varied knowledge base. Their contributions to science, and to what I label the *expansionary* vision, are immense.

Vladimir Vernadsky, geologist, was a leading light of the *Russian cosmism* school. He invented the concept of the *biosphere* in 1917, and in 1924, as we learned, introduced the notion of the *noosphere*. For Vernadsky, the noosphere concept represents the sphere of intellectual life. His concept of the noosphere differed from that of Teilhard and many of the modern futurists, who see the noosphere as strictly earthbound, a ring encircling our own planet. Vernadsky's noosphere, on the other hand, corresponds to its meaning in the expansionary vision of the human future. In his theory humankind would eventually expand the noosphere—human consciousness and intelligence—into nearby space and eventually into the farther reaches of the solar system and beyond. This process closely reflects this volume's concept of vitalization, humanity projecting into the universe both his physical and spiritual presence.[1]

Other contributors to Russian cosmism parallel expansionary thinking. Vladimir Odoevsky, for instance, predicted that humankind would travel to the moon to mine lunar mineral resources for the benefit of Earth's population. In the mid-nineteenth century, Vladimir Solov'ev claimed that humans would progress to a state where they would be able to exercise an almost godlike control over the universe. One late nineteenth-century proponent of cosmism, Nicholai Federov, theorized that humans, as beings of the highest consciousness, were obligated to introduce design and purpose into the chaotic workings of the natural world. Federov predicted that humankind's need for living space in which to materially and spiritually develop would compel us to eventually settle the planets in our solar system. Federov was mentor to Konstantin E. Tsiolkovsky, and he instilled in his protégé a desire to make human space flight a reality. By the end of his life in 1935, Tsiolkovsky had made seminal contributions to the theory of reactive propulsion, developed the fundamental principles of liquid propellant engines, and originated such ideas as orbiting space stations, transparent domed habitats on the surface of asteroids, and multistage rockets.

Cosmism's strength emanated from its emphasis on what we would now probably refer to as a multidisciplined approach to knowledge. Cosmism expected its adherents to be "Renaissance men." Long before Tsiolkovsky had developed his scientific skills, he was already speculating about issues such as the origin of the universe, the high moral aspects of human life, and the purpose of human existence. His early philosophic inquiries served as a catalyst for his scientific achievements. First, he decided that humankind must travel into space to meet its destiny; then he embarked on a career that would establish the scientific principles that would make space flight a reality.

In fact, we see time and again this principle borne out. The very act of visioning provides the energy and motivation to unlock the secrets of nature. Ideas precede reality! Conversely, a society that has forgotten how to dream will no longer create a worthwhile future. No vision, no progress!*

Tsiolkovsky (who incidentally has a Moon crater named after him) wrote a seminal work in 1932 entitled *Cosmic Philosophy.* In this work he combined religion, morality, space science, and human destiny into a utopian vision of the future, one in which science and technology are harnessed to attain universal happiness. We can see the influence of this eclectic approach on the work of Hermann Oberth, the Rumanian-born rocketry pioneer. Oberth's *Man in Space* (1954) deals with both technical aspects of space flight and the philosophic aspects of human colonization of deep space.[2]

Cosmism clearly influenced modern Western thinkers. Robert Esnault-Pelterie, who lived in the first part of the twentieth century, was a giant in the field of modern astronautical theory. Like the cosmists, he felt that space exploration's ultimate value would be to help us understand the genesis of life and human progress. The modern vision of large-scale space settlements reflects the deep tradition of cosmism philosophy. For instance, Freeman Dyson, whose ideas we will soon discuss, introduced the idea that we may colonize space by building so-called Dyson spheres for human habitation around stars. In the 1970s, Gerard K. O'Neill stated that large city-sized space stations might be the optimal living environments for high-technology civilizations.

Visions for the Twenty-First Century

As we begin the third millennium, ideas and theories that I feel are reflective of the expansionary point of view are proliferating. These writers, scientists, cosmologists, and futurists develop the new expansionary theories using modern statistical analytical tools, probability theory, mathematical modeling, and computer simulations to formulate and test their theories. In addition,

---

* Most of the Cosmists had the time, or made the time, to contemplate the big picture and "play" with ideas. Free time, time not allotted to career or personal responsibilities, seems to be the wellspring of many of the great discoveries, inventions, and original concepts that move society forward. Litterateur Russell Kirk and others have gone so far as to proclaim that leisure is the basis of culture.

the Internet enables researchers to easily and quickly access each other's work and share perspectives.

One of the currently less widely known futurists whom I consider expansionary in his thinking goes by the conveniently nearly anagrammed name F. M. Esfandiary. (He later changed it simply to "FM-2030," combining his initials and the year 2030 to "reflect his beliefs and confidence in the future.") At the time of their publication, his books, sporting such titles as *Up-Wingers* and *Are You Transhuman?* received accolades such as "daringly original" by the *Village Voice, Omni* magazine, and the *Washington Post*. Futurist Alvin Toffler considered Esfandiary "an exhilarating voice of a new, nonmystical consciousness."[3] He saw humans playing a major part in the future development of the universe, considered space colonization a natural evolutionary step for the species, and felt that by increasing their life span humans would be better situated to develop this and other planets.[4]

Nobel laureate Freeman Dyson certainly fits into the expansionary mode as defined in this book. In fact, many of his scientific and philosophical ideas reflect the spirit of the Russian cosmism school. In a recent book, *Imagined Worlds*, this visionary thinker audaciously projects a million years into the future to ascertain the fate of the human species. According to Dyson, meaningful space travel and colonization, especially of the moon and Mars, is still about 100 years away. Once we do start moving into space, however, millions of people will become involved in the colonization effort. A thousand years hence our species will be inhabiting human settlements throughout the solar system, as far out as the Kuiper belt of comets, a thousand times the distance between the Earth and the sun.

According to Dyson, by that time humankind may have already undergone significant *speciation*, the division of our species into many variants. Such genetic differences between human species may be the result of natural adaptation to totally alien environments, or our more proactive efforts such as genetic engineering. The population of the human species may be 500 times its current size. Ten thousand years from now the human species will have achieved a form of immortality and will have mastered most of the physical laws of the universe. One hundred thousand years from now the human species will have spread throughout the entire galaxy. Dyson tempers his optimism with a concern that humans living at such distances from each other will lose touch with each other. I would counter that communications technology should be so sophisticated that we will be able to transmit signals quite easily across the galaxies. (I will offer what I believe is a unique solution to this communication problem in Chapter 7.)

Ironically, his most optimistic forecast about the fate of the human species lies not in the near term but at least one million years from now. By this time, the human species, whatever physical form it will have assumed, will begin to meet the mandate envisioned in this book.

> A million years from now, our descendents and their neighbors in the galaxies will perhaps be preparing for the intelligent intervention of life in the evolution of the universe as a whole.[5]

Such a statement represents expansionary thinking at it most creative and most daring!

Physicist Michio Kaku, cofounder of string theory, and author of two well-known books on physics and the future, clearly belongs on this select list of twenty-first-century visionaries. Like Dyson, Michio Kaku takes the long view of human development clearly influenced by expansionary Russian cosmism school. According to Kaku, our species is making a transition from passive bystanders to active choreographers of nature. Over the next century or so, we will eliminate genetic disease by injecting people's cells with the corrected versions of genes. We will be able to eliminate large classes of cancers without invasive surgery or chemotherapy and grow entire organs in the laboratory. Kaku looks to a "second industrial revolution" (which I contend throughout my work has already begun) that will enable us to use nanotechnology to conquer scarcity and poverty.

One of Kaku's major contributions is his revival of the ideas of Russian astronomer Nikolai Karashev, a visionary who devised a useful typology to classify societies. Karashev envisioned technological civilizations advancing through several phases, which he labeled Types I, II, and III. Type I refers to a global civilization, the masters of a single planet. A Type II civilization utilizes the resources of an entire solar system; such a society might for example build a vast shell or Dyson sphere around its star. A Type III civilization operates on a galactic scale, occupying numerous solar systems. Within this schema, humanity at its current stage of development is a backward society, or Type 0, but is on the verge of attaining Type I status. Kaku posits that humankind will take many centuries to reach Type II and many millennia to achieve the galactic Type III status. Nevertheless, it will be well worth the effort: Civilizations of Types II and III have so mastered the forces of nature that they have become invulnerable to asteroid impacts, supernova explosions, and other natural disasters.

Karashev's expansionary vision sees the species progressing toward greater control of itself, its environment, and eventually the universe. From Karashev's point of view, the human species brings value to the universe—we advance ourselves not just for humanity's aggrandizement but for the universe's progress as well.[6]

Two other twenty-first-century figures, both visionary/activists, should also be included in the ranks of the expansionary vanguard. We met Robert Zubrin earlier, and learned how he is directly impacting the NASA Mars Exploration projects. His Mars Direct concept is already attracting a growing allegiance of adherents. He and his organization are moving quickly from theory to action—they plan to build a prototype space base in the Arctic and launch their own spacecraft in the near future. In addition, Marshall Savage, in *The Millennial Project*, presents a cogent strategy for colonizing the solar system. He exhibits the expansionary philosophy's positive attitude toward humankind and the central role it plays in the universe's future. Moreover, Savage spends a good deal of time educating the public on technological issues through his Living Universe Foundation. He also toils tirelessly raising funds to build a prototype space colony, possibly an artificial island in the middle of one of our oceans.[7]

As we enter the twenty-first century, interest in the concept of human destiny is accelerating. January 2000 saw the publication of journalist Robert Wright's book, *Non Zero: The Logic of Human Destiny*, as an attempt to explain human history from the perspective of anthropology and game theory. Wright deduces that human history is moving in a positive direction, toward what he labels higher complexity—on both the physical and social levels. The book can be categorized as expansionary in that it emphasizes the inherent value of the human being and portrays our species as one primed for greatness.[8] However, according to the majority of critics, most who praised the book's breadth and scholarship, Wright does not deliver on his subtitle. A major *New York Times* review of the book complained that "For all its boldness and chutzpah, *Non Zero* suffers from a failure of nerve" in that "the implications of the teleological program are not carried to their logical conclusion." In short, Wright presents no clear vision of the nature of human "destiny."[9]

The fact that so many critics complained about Wright's omission of a clearly defined concept of human destiny in itself strongly demonstrates just how intensely our society hankers for such a vision. Humans require a sense of purpose and are sorely disappointed when one is promised but not delivered.

Let us see if the expansionary vision of human development can answer the questions Wright and others leave unanswered.

Emergence of the Expansionary Vision

As we can see, new visions, comparable in spirit and oftentimes substance to the vision encapsulated in this work, have been emerging over the last century. These theories, representing the emerging expansionary vision of the human species, have the following commonalties, with some variations.

One common thread running through all these philosophies is the contention that the human being has a purpose, even a destiny. Importantly, these new theories repudiate the misanthropic notions bandied about in the late twentieth century, especially the extreme idea that somehow the world would be a better place without our presence. Quite to the contrary, many of the visions considered herein proclaim that humankind is crucial to the universe's future development.

Each of these philosophies plainly states that the human species is unique, not just a continuation from another species or just a smarter version of a lower species. They portray humankind as qualitatively different from whatever preceded our appearance on the planet.

All these philosophies also state that the human species will have to transform itself into a spacefaring society, for its own growth and that of the universe. How else can we have a maximum impact on the universe unless we go there? In addition, a critical component of this emerging expansionary vision is the belief that out of sheer necessity the human species will evolve further, perhaps into several branches scattered throughout the universe.

A central tenet of this emerging vision of humankind is the belief that the human species is so unique that only we have the ability to transform the universe. Hans Moravec and Ray Kurzweil's insistence that robots or so-called smart machines will either replace or absorb the human species put them in a position contrary to the expansionary framework. These machines, regardless of their advanced computational and physical capabilities, are simply not human! They can never possess or express the creativity, the serendipity, the purely transcendental qualities engendered in human consciousness. Therefore, the universe would gain little from the dispersal of these nonhuman entities into the cosmos.

Where do these expansionary visions diverge? Ironically, the Russian cosmism school's vision of the human species, which originated over a century ago, seems closest to the expansionary model. Of the contemporary thinkers,

Savage and Zubrin engender more of the spirit of this philosophy than the others—their visions fairly sparkle with the pioneering spirit; they are adventuresome, curious, and anxious to make the future happen. Although Dyson and Kaku both unabashedly embrace technology and science, they both occasionally betray suspicions that the human species may misuse it. They seem overly worried about problems such as overpopulation, pollution, and, in Dyson's case, man's "warlike" tendencies. At times, I am left wondering why they would want a species with such imperfections to transmigrate across the galaxies.

In spite of their differences, these theories converge around one core principle: humankind has a special role to play and has appeared in this cosmos for a reason. But how do we know such a theory contains even the slightest fragment of truth? Where is the proof for this seeming exaggerated claim for the worth of the human species?

These claims, however inspiring, must be grounded in science. In addition, science would ask many questions before it would even begin to entertain the idea that the human species is imbued with a special "destiny." For example, if humanity were necessary, were conditions arranged to make its appearance possible? What proof do we have that humanity was anticipated or planned for? Moreover, even if circumstances did conspire to arrange the emergence of our species, the next question is why? What is this mandate for which humankind seems to be preparing itself? And did the human species aid and abet his own evolution to put himself in position to meet this mandate? In addition, why should humankind want to achieve this destiny in the first place? Finally, is all of this coming to a head in the twenty-first century as we advance our intellectual and physical abilities at such a breakneck pace?

Let us begin to explore these issues and try to answer some of these questions.

The Universe Beckons: The Coming Dominance of Anthropic Cosmology

Throughout the book I have claimed that the human species possesses a unique destiny, and have alluded to the nature of that destiny. This implies that humankind has been anticipated, planned for, summoned, and most definitely desired in the universe.

One condition that I believe bolsters such a belief is the growing evidence gathered over the last few decades that the emergence of life was facilitated by the existence of a unique set of physical conditions existing throughout the universe. If the calibrations for these physical forces, including gravity and electromagnetism, were even slightly different, the human species could not have emerged. Recent evidence strongly suggests that the appearance of human life was hardly an accidental occurrence.

The public has basically accepted the notion that our world, and our species, is a result of entirely random events. As you will see, quite the opposite is true. According to a novel new theory, the *anthropic principle*, of all the various ways the universe could have formed, and the various values it could have given to such forces as gravity, the universe came forth in just the correct way so that life and life forms could exist.

The novel argument embodied in the anthropic principle has its foundations in a paper presented in the early 1970s at a conference to celebrate the 500th birthday of the father of modern astronomy, Nicolaus Copernicus. Attending this conference were such scientific notables as Stephen Hawking, Roger Penrose, and Joseph Silk. At this conference speakers presented papers dealing with the latest astronomical discoveries and cosmological speculations. Among all the papers presented, one would survive to become the most revolutionary presentation of the entire conference. Brandon Carter, an astrophysicist and cosmologist from Cambridge University, introduced this earth-shattering concept in a paper titled innocently enough, "Large Number Coincidences and the Anthropic Principle in Cosmology."

What was curious about his paper, of course, was the term *anthropic principle*. He named his notion after the Greek word for man, *anthropos*. His findings challenged the idea held by scientists since the nineteenth century that we live in a random universe in which humans appeared essentially by accident, through material forces haphazardly unfolding over the eons.

Before describing the implications of this theory, let me explain its rudiments. The anthropic principle plainly states that all the seemingly arbitrary and unrelated constants in the physical and cosmological world have one thing in common: they all have the values that are needed to produce life. To put it more starkly, from the very beginning of the universe the myriad laws of physics were compiled or situated in such a way that life—no, a carbon-based entity like humanity—could emerge. The laws of physics appear to be expressly designed for the emergence of human beings.

Carter's principle offered an explanation for one of the most basic mysteries of physics. For decades physicists have been unable to explain why the value of so-called fundamental constants, such as the values of the gravitational or electromagnetic forces, are what they are.

Carter claims that these precise values and ratios were just those values necessary for the universe to be capable of producing life. Or to rephrase Carter's claim, if the universe intended to produce life, *these precise values or ratios had to exist.* There were just too many values that had been arranged around the central task of producing us for the whole scheme to be a set of random events. According to Carter, these facts suggest that humankind's position in the universe, if not "central," is "privileged." Without even approaching the reasons why we might be the objects of the universe's affection, let's look at some of the strange coincidences Carter and others have cited that have conspired to produce life, and eventually humankind.[10]

Much of the anthropic principle is tied to the Big Bang theory, which claims that roughly 15 billion years ago all the matter and energy of the universe, and space and time as well, were concentrated in a superhot region, a fireball known as the Big Bang. According to this theory the universe began with an initial explosion of this dense, hot soup of subatomic particles and radiation. For thousands of years after the Big Bang, radiation did not stream freely into space, but was repeatedly absorbed and scattered by these charged particles. About 300,000 years after the Big Bang (a blink of an eye in relative cosmological terms), the universe became cool enough for the electrons to combine with nuclei. (We supposedly even have a snapshot of this "background radiation," the universe as it was shortly after the Big Bang.)

Our current observations tell us that other galaxies are rapidly rushing away from ours. It seems as though the universe is constantly expanding—all spheres, star systems, and galaxies are pushing out through the cosmos all because of that initial explosion billions of years ago.

The theory has its shortcomings, to be sure. We still cannot determine how this initial explosion of particles led to the formation of galaxies, planets, stars, and suns that we see now, including the planet we live on. Moreover, new evidence is always turning up to upset our understanding of the mechanics of the Big Bang. For instance, up until a year or two ago we thought that the universe's expansion was actually decelerating. Recent images sent back by the Hubble Telescope and other instruments have convinced many observers that the universe is in fact moving apart at an ever-quicker pace.[11]

Scientific investigations of the Big Bang theory revealed a number of conundrums. Physicists tried to construct numerous alternative scenarios of the universe's evolution. For instance, in computer simulations of the evolution of the universe they tinkered with the value of gravity as it exists in our universe; they altered very slightly the strength of the electromagnetic force in the cosmos. Every time they changed these values, they found that the universe would not evolve as it actually had. In some scenarios, for instance, if you changed the value of gravity or electromagnetism, making it stronger or weaker than it actually is, you ended up with the wrong kind of stars, or no stars at all. In other words, the values for gravity and electromagnetism had to be at precisely specific levels for the universe to look the way it does today.[12]

According to Timothy Ferris, in his wonderful compendium *The Whole Shebang*, the conditions in our universe are *perfect* for the emergence of life. Gravity, for instance, is at the optimum level for the expansion of the universe. If gravity was any stronger a force, cosmic expansion would have halted and the universe would have collapsed long before life, and humanity, could have evolved anywhere. Even if expansion somehow continued, stars would burn out too rapidly to incubate intelligent life on the terrestrial time scale. The sun would have lasted only about a billion years, and the planets might never have emerged. In short, if gravity did not possess just about exactly the strength it does, humankind would not have ever walked forth on the cosmological stage.

If any of the other physical constants of the universe were of different strengths, life as we know it could not have emerged. These conditions would produce universes with stars unlike our sun, too cool or too hot to sustain life's evolution. Ferris posits other conditions in which the universe would consist of hydrogen but no other elements—meaning no oxygen and hence no human species. In some cases, different gravitational strengths would have resulted in a universe entirely composed of helium or a universe without protons or atoms. If gravity was stronger than it actually was, the universe would have collapsed back in upon itself before the first moments of its existence had transpired.

The anthropic principle permeates the physics of the universe—the basic forces of the universe are such that they are *perfect* for the emergence of humans! If any of those constants had been slightly different, stars would have burned out much faster than would have been necessary to support life; the existence of water would not have been possible. And not just "any" water, but water which mysteriously is lighter in its solid than liquid form, a lucky occurrence which permits ice, the solid, to float on the liquid form of water. In a world where ice is heavier than liquid water, oceans would freeze from the bottom up and Earth would now be covered with solid ice. Life and humankind would never have evolved in this world!

Physicists have been observing for several decades that these "cosmic coincidences," a term coined by John Gribbin and Martin Rees in their book of the same name, are pervasive. The constants of the universe fall within an exceedingly narrow band that is compatible with life. Some, like Freeman Dyson of the Institute for Advanced Study, ask whether life therefore is a special property of the universe. Dyson has exclaimed, "It's as if the universe was expecting us."[13]

Generally, the scientific establishment has reacted to the emergence and dissemination of this novel anthropic principle with mixed emotions. Physicists have plainly found the existence of such a principle downright annoying. They have always suspected that any theory that contends, or worse proves, that the universe anticipated life, was created with life in mind, smacks of "intelligent design" theory.

The physicists and cosmologists real discomfort stems mainly from the fear that to apply *teleological* metaphors to the universe, to claim it has a purpose, sounds more like religion or metaphysics than science. And truth be told, the "revelation" that the universe's physical qualities so closely match those required for life, human beings, to emerge, survive, and thrive have prompted theologians and religious laypersons to surmise that an "intelligent designer" must have been pulling the strings, manipulating matter, gravity, and other forces to ensure that the human race emerged.

Thus science, instead of embracing or trying to make sense of the anthropic principle, has spent two decades developing a wide range of brand new theories to dispel the anthropic principle. Ironically, some of these alternative theories themselves sound downright religious and mystical. For example, in their efforts to refute this principle some modern cosmologists have developed the notion that the universe we live in is only one of countless billions of universes that coexist simultaneously with ours. They claim we are living not in a universe but a "multiverse."

The argument the multiverse camp uses to debunk the solid math, physics, chemistry, and pure science and logic underlying the conclusions of the anthropic principle goes as follows. It seems undeniable that the universe did in some way make special physical preparations for the emergence of life, and might have even summoned forth life, and humanity. However, they claim, there are billions of other universes, existing parallel to the one we live in, in which conditions would not allow humanity to emerge. Those sympathetic to this vision of a multiverse do not seem bothered by the fact that we cannot see, measure, visit, or send probes to these alternate universes.

According to Kaku, "Perhaps there are an infinite number of possible universes, with different physical constants. We just happen to live in the one that is compatible with life." In Kaku's multiverse, or world of parallel universes, "it is not an accident at all that the physical constants are compatible with life; we coexist with plenty of dead universes where the physical constants are incompatible with stable DNA-type molecules." These other universes, it is assumed, have totally different physical and molecular laws, most of which would not allow for life, and especially human life. Lucky for us, we happen to live in the universe that presupposes life.[14]

The multiverse concept has met with both awe and derision. One of the original proponents of the multiple universes concept, American scientist John Wheeler, eventually abandoned the existence of infinite invisible universes because he considered this belief a product of conjecture that borders on a kind of mysticism. On the other hand, other cosmologists such as Paul Davies have written that they and others find the idea of parallel universes superior to the idea of supernatural design. Timothy Ferris seems to want to cut a deal with the cosmos. He will accept the possibility that the anthropic principle is valid, under the condition that we also "entertain the hypothesis that there are many universes, each with its own set of physical laws."[15]

The more I am exposed to the evidence behind the anthropic principle, the more I am convinced that the universe established the conditions for the emergence of intelligent life. Moreover, the more I study the recently devised alternatives to the anthropic principle, the more I am convinced that these counter-theories—multiverses, for instance—are simply less plausible than the anthropic principle. In addition, none of these "improvements" are as yet testable or scientifically refutable. In fact, the physicists are now sounding more like the religionists of old, resorting to arguments based on faith and speculation (and perhaps their secret yearnings); whereas the supporters of the anthropic principle defend their position using the basic principles of physics and chemistry.

At this point, unless we are willing to believe some particularly arcane, fantastic, and possibly unprovable theories, we have no choice but to accept that fact that the universe we know, the one being photographed by Hubble and explored by Cassini, is the only universe we have. Moreover, this is the universe uniquely conditioned to produce life. This position is supported by a phalanx of facts that stand on their own—all the factors we reviewed heretofore.

In fact, it is hardly a stretch to claim that not only did the universe establish the conditions for life—it actually summoned life up! The universe seems to have manipulated the laws of nature to ensure that life, and possibly humanity, would come to fruition.

Such a state of affairs does not require or presuppose, as many cosmologists fear, the existence of a supernatural intelligent designer who choreographed the dance of protons, neutrinos, and life itself after the Big Bang. That being said, let me also strongly suggest that physics seems to support the position that the universe does have reason to have situated in its midst a sapient creature, one with consciousness, intelligence, and creativity.

Humankind's Purposive Self-Development

The anthropic principle may explain the establishment of conditions enabling life to emerge. However, this principle is not sufficient to explain the appearance of a species as sophisticated as the human being. To even attempt to gain an understanding of the emergence of the human species, we must consider the claims of the theory of evolution.

On its face, evolutionary theory does not seem particularly congruent with an expansionary vision of human development. The theory propounded in this book portrays humankind as a creature endowed with a resplendent destiny. Modern Darwinian theory, on the other hand, sees human evolution as a randomized process largely devoid of anything remotely resembling purpose, direction, let alone destiny.

In the modern evolutionary view, humankind's appearance on the planet was basically the result of an incredible streak of luck. And of course to even entertain the idea that humanity possesses a destiny or purpose is considered nonsense, given the randomness in our very appearance on the planet. To quote Harvard biologist Stephen Gould, Darwinism proves that humans are merely "a fortuitous cosmic afterthought, a tiny little twig on the enormously arborescent bush of life."[16]

Let's listen to what some modern proponents of Darwinian evolution have to say about humankind's destiny. In 1967, paleontologist George Gaylord Simpson, in his book *The Meaning of Evolution*, wrote of our evolution thusly: "Man is the result of a purposeless and natural process that did not have him in mind." In 1970, molecular biologist and Nobel laureate Jacques Monod announced that now that we accept the scientific underpinnings of

Darwinism, "man has to understand that he is a mere accident." And in 1986, zoologist Richard Dawkins's best-selling book, *The Blind Watchmaker*, possessed a subtitle that reveals his position on the matter: *Why the Evidence of Evolution Reveals a Universe without Design*. In another work, *The Selfish Gene*, Dawkins went so far as to claim that the real reason for any organism's existence is to carry genes who are looking to replicate themselves. The gene's host helps the gene in its quest by engaging in such processes as insemination, impregnation, and birth. Once the carrier does his or her job, the gene has little further use for its carrier. Unsurprisingly, the Oxford zoologist has written that the universe "has precisely the properties we should expect if there is, at bottom, no design, no purpose, no evil and no good, nothing but pointless indifference."[17]

Obviously, we make claims in this book which directly contravene such judgments of our species. The expansionary view declares that the human species has been developing along its unique evolutionary track toward a being possessing a complex physical structure, sophisticated intellectual capacity, and most importantly consciousness. Can the principles of Darwinian evolution be reconciled with the vision of the human species as one with a direction, purpose, and destiny? The answer is *yes*. However, some wrinkles must be added to the theory, and some newly emerging scientific syntheses and findings considered, for evolution and the current expansionary vision to work together.

The human species has developed into an extremely sophisticated and intelligent being in what seems a relatively short geological time frame. It would seem that such rapid growth in our development required a certain degree of *intentionality* that most Darwinian theorists refuse to acknowledge. That is, we will have to posit that the evolutionary process is guided by at least a semblance of consciousness and purpose on the part of the actors, and reject the neo-Darwinian belief that evolution is an inherently random process. I will later offer a concept I label *purposive self-development*, which I believe is and has been an underlying *motif* of human evolution since the species first started evolving.

Darwin's Breakthrough

In his meticulously researched book *The Origin of Species*, Charles Darwin described in great detail a process by which all plants and animals have gradually evolved from primitive life forms by the natural selection of random variations. This process has taken vast periods of time, at least hundreds of millions of years.

Let us quickly review Darwin's ideas, some of which we introduced in our earlier treatment of biogenesis. According to Darwin, just as domestic livestock can be modified selecting certain variants for breeding, so species in the wild can be modified by a natural selection due to competition for survival. Those with the traits best suited for survival in that particular environment—e.g., strong teeth, ability to see in color, large physique—will be more likely to live long enough to breed, and breed plentifully, than would be those without such survivability traits. The survivor's characteristics will be passed on, and if these individuals produce more offspring than those with the less favorable characteristics produce, over time their characteristics will become the predominant traits of the species. As more genetic mutations are introduced into and dominate the species gene pool, species change, that is, they *evolve*, into new variations of the species; sometimes the species experiences variations so radical over a period of a million years or more that its descendants can be considered a wholly new species.

Darwin used as proof of this process the fossil record that showed species evolving over time, adding some traits, losing others. Darwin claimed that all living forms could be described as being part of the same family tree. That is, the continuation of such "descent with modification" over millions, even billions of years produced all living things from one or a few original organisms. He saw no direction or purpose in this process—it was completely random, all aimed at survival of the species. No design was involved.

Darwin's theory, on the face of it, seems completely logical. After all, a species would desire to retain positive traits, such as functional eyes and upright posture, and discard the undesirable traits. Darwin's theory requires that the desired traits be inherited by the offspring of the selected individuals. How, though, does this process actually happen?

Genetic research that emerged in the 1930s and 1940s provided the clue to how these traits would be passed on by a species. This modern synthesis of Darwinian theory and genetics states that all organisms are composed of traits that are largely genetically determined. A gene exists for each trait, for hair color, eye color, and the like. If a trait confers an advantage in the competition for survival, the gene for that trait is more likely than others to be transmitted to the next generation. New traits often appear because of mutations in the DNA of members of the species. The modern Darwinians conclude that over the course of many generations, the advantageous gene, and of course its corresponding trait, will be found in a higher proportion of individuals in that population. Such a process can take centuries, eons perhaps, a time span that has become a major source of criticism of Darwinian evolution.

In other words, this theory would suggest that the species is wending its way through the centuries, basically changing through random trial and error, mutation, and response to changes in environment, geography, and climate.

The concept of change over time and the process by which it happens seems logical and difficult to refute. Yet, a barrage of criticism has been directed at this theory. For one thing, many do not think that the details of a complex organism's structure and physiology could have been formed so exquisitely and perfectly through the random workings of evolution. Others doubt that, given the amount of time for the process to occur, life could have evolved into its current state simply through a randomized process of natural selection.

Physicist Gerald R. Schroeder, a former MIT professor now with the Weizmann Institute in Israel, through painstaking statistical analysis, has concluded that evolution, at least the way we understand it, could not have occurred randomly. The process of species waiting for advantageous traits to appear through DNA mutation and then selecting for these traits through breeding seems arduous, complex, and hence incredibly time-consuming. Schroeder claims that probability theory strongly suggests that in the time spans available, say 5 to 7 million years, species would have had a hard time developing into their current complex states if they did so strictly through such random trial-and-error methods.

His most stunning finding is how mathematically improbable it would be for the human species to evolve from its nonhuman ancestors over a period of 5 million years by *random* natural selection. According to current biological and paleontological theory, about 7 million years ago, in Africa, a divergence occurred from a common ancestor. Two separate species, the chimpanzees and Homo sapiens emerged. About 4 million years ago the *Australopithecus*, appeared, an upright forerunner of the human being. *Homo habilis* ("Handy Human") appeared between 2.3 and 1.3 million years ago and showed greater growth of the brain and increased use of tools. *Homo erectus* appeared about 1.5 million years ago, and began evolving into *Homo sapiens* only about 300,000 to 500,000 years ago. The Neanderthal was the first version of *Homo sapiens*. About 50,000 years ago, the *Cro-Magnon* man appeared. Only around 15,000 years ago did the fully developed human being appear. Throughout this process, our species got taller; our ability to produce sounds, such as language, improved dramatically; and more importantly, our brain grew progressively larger.

Schroeder emphatically asserts that the neo-Darwinist position cannot explain how we evolved to human from a chimpanzee-like state in that period of time. To get from this very primitive physical state to *Cro-Magnon* man, millions of mutations would have had to occur. According to the calculations of Schroeder and others, mere randomness would never get us to where we are today from this chimp ancestor in the time allotted. As he says, "the problem of producing us by the random plus selective natural process becomes insurmountable in the time available."[18]

How improbable? Even if we had mutation rates a *million times higher* than currently observed in genetic behavior, we would require an astounding *one hundred million generations* for the chimp to reach the human stage. If we assume more normal mutation rates, the required time reaches hundreds of billions of generations.[19]

Schroeder claims that confirmed evolutionists are aware of the mathematical improbability that through random selection we would not have had enough time on the planet to evolve to our current state. Such troubling statistical improbabilities regarding the evolution of humans were revealed at a famous symposium at the Wisteria Institute in 1967. This symposium had brought together biologists and mathematicians to find a mathematically reasonable basis for the assumption that random mutations are the driving force behind evolution. Although the mathematicians revealed to the biologists the daunting statistical improbabilities of evolution by random mutation, the biologists at the conference stubbornly managed to retain their faith in evolution by random mutations. They did so by deducing that since the theory of natural selection is true, *the mathematics used to refute random selection must somehow be faulty.*

By clinging to the concept of randomness, Darwinian evolutionists have unnecessarily allowed themselves to be subjected to criticism from mathematicians and other scientists. Ilya Prigogine, a Nobel prize winner in chemistry, claimed in an article in *Science* that it would be almost impossible for a macroscopic number of molecules to give rise to the "coordinated functions characterizing living organisms." He was quite clear: "The idea of the spontaneous genesis of life in its present form is therefore improbable, even on the scale of billions of years." In other words, there was just not enough time for us to evolve.[20]

Suggestions abound over how we may explain the appearance of humankind without resorting to the Darwinian concept of random natural selection.

Happily, such new paradigms are friendly to the inclusion of concepts such as purpose and direction within the evolutionary framework.

Self-Organization and the Challenge to Randomness

One of the problems we have in comprehending how species can progress from simple to complex, from chimp to *Homo sapiens*, in what many consider unreasonably short times is our understanding of randomness and the role that it plays in this process. In short, we will have to radically reconsider just how much weight to give random actions in evolution's schema if the theory of evolution is to successfully overcome some of the internal contradictions modern critics have exposed.

New findings in fields such as physics and biology may explain more clearly how systems actually evolve in specifically nonrandom ways. There is mounting evidence that systems evolve from simple to complex with some degree of *intentionality*; that is, the actors within such systems, whether they be molecules or animals, act according to some innate inclination to seek order and growth. To understand such a concept, let us first look at the behavior not of humans but of molecules.

New discoveries in physics and biology strongly suggest that when given the opportunity, life will emerge from its surrounding environment and the materials existing therein. For instance, under the right conditions even simple molecules will rearrange themselves into an orderly sequence that will make the emergence of life possible.

One of the key questions facing evolutionary theory is how lower, unicellular forms evolved into higher, complex organisms. In an article in *Science News*, biologist Richard Lipkin claims that new research strongly implies that the origin of life was not random at all, but was instead a highly ordered, determined, chemical phenomenon. If the research he reports is replicated, it would imply that evolution, and by implication the appearance of life, and humans, was not random or accidental. Instead, the chemical steps that turned nonliving molecules into living cells were actually constrained by the molecules involved and their inherent tendencies to aggregate in specific ways.[21]

In 1953, Miller and Urey, two biochemists at the University of Chicago, showed that amino acids could form out of complex molecules under primordial conditions. Subsequent research has demonstrated that amino acids could assemble themselves into simple proteins, which serve as the core structural components of living cells. These proteins could fashion themselves into cell-like objects called protein *microspheres*. These are "empty cells"— they don't have the internal machinery that runs a living cell, but can create a network, joining together and signaling each other electrically when stimulated by light.

Over the years, this controversial concept, that dead inorganic cells can evolve into living cells in an ordered, determined sequence, has gathered many proponents. Aristotel Pappelis, a biologist at Southern Illinois University at Carbondale, supports the idea that proteins might have formed these microspheres that became precursors of modern cells. Importantly, he makes this statement: "This is a *highly determined sequence of events* that occurred on Earth, gave rise to life, and made evolution possible." In other words, there seems to be some form of purposive behavior exhibited at all levels of material existence.[22]

Among those working on the origins question is biologist and Nobel laureate Christian de Duve, who has outlined a theory of how life might have arisen. He dubs his theory the "thioester-iron world," after the chemicals he thinks could have reacted together to create "protometabolisms" that could evolve. He admits his theory is very speculative, but believes that one day biologists may find traces of the prior existence of these protometabolisms in the biochemistry of contemporary organisms.

Some claim that such tendencies for the evolution of life, of material, of molecules, to occur in nonrandom ways is the result of laws whose existence we are only now beginning to suspect. These laws of complexity spontaneously generate much of the order of the natural world. Scientists at the Santa Fe Institute studying complexity theory argue that the laws of physics and chemistry make the emergence of life, given the presence of certain materials in hospitable environments, practically inevitable.

In his book *At Home in the Universe*, Santa Fe Institute biologist Stuart Kauffman argues that the emerging science of complexity suggests that the order we see around us is not all accidental, or random. As he says, "Profound order is being discovered in large, complex, and apparently random systems." He goes on to say that "this emerging order underlies not only the origin of life itself, but much of the order seen in organisms today." As he claims, this would lead to a revision of the Darwinian world view of the emergence of man: "Not we the accidental, but we the expected."[23]

Like the aforementioned scientists, Kauffman also believes that life emerged through the reactions of chemicals that were in the primordial soup. Kauffman postulates that if there are enough different types of compounds in the chemical soup, they will begin to act in metabolic ways and be able to reproduce and evolve.[24]

Increasingly, many people are beginning to recognize the tendency of matter to self-organize, demonstrating a powerful inclination in the universe toward order when the conditions are right. Moreover, some noted scientists are beginning to think that some inclination or tendency toward order had to be in existence if man was ever to appear. Christian de Duve states that "eventually we will understand that the origin of life was not a highly improbable cosmic jest but rather an almost obligatory outcome of chemical structures, given the right conditions." Ian Stewart, a mathematician at the University of Warwick in the United Kingdom, has suggested that mathematical rule structures inherent in existence may eventually be shown to operate as if life were their goal, encouraging the development of animation. "DNA may be just one of the many secrets of life, secrets we are only beginning to glimpse," Stewart suggests.[25] Such statements seem to reflect the anthropic principle, but now on the biological level.

Is it possible, then, to find not a middle ground but a third way to understand the evolution and development of the human species that does not depend on either randomized genetically based natural selection or concepts of intentional design? I believe there is, but I think we must recalibrate our thinking about the purpose of man, and therefore of evolution.

A Species with the Best Intentions

The emerging picture of early Earth is one of a planet brimming with activity, virtually forcing life into existence. As soon as the molecules had the chance, they attempted to establish the conditions for life. This self-organization of molecules made life and the evolution of life forms possible.

It is the contention here that the same inclination to self-organize, to intentionally evolve oneself from the simple to the complex, exists on the biological level as well as the molecular. And the human species is the finest example of this process.

Alfred Russell Wallace, a contemporary of Darwin who concurrently developed a similar theory of natural selection, discussed a major mystery in human evolution. It seems that between *Homo habilis* and *Homo erectus* the human brain underwent a gigantic jump in its size. The earlier hominid had a brain only slightly larger than that of an ape. *Homo erectus*, which existed for a million years starting around 1.5 million years ago, had a cortex as large as ours. Wallace contends that the human brain was overdesigned for its primitive uses and thus could not have been a production of natural selection. He said that "natural selection could only have endowed savage man with a brain a few degrees superior to that of an ape, whereas he actually possesses one very little inferior to that of a philosopher."[26]

Robert Orenstein, a biologist specializing in brain research, is similarly curious about why *Homo erectus* possessed a brain that he ostensibly had little use for. Our brain expanded to a size for which there was little functional use at the time. According to Orenstein, *Homo erectus's* brain was complex enough to invent a microprocessor, even though all that was needed at the time was a brain that could figure out how to hammer out the first few stone tools. "Why be able to fly to the moon when no one has even understood how to make iron?" Orenstein asks.[27]

Why indeed be equipped with a brain able to build spaceships when we are just learning how to build a fire? The early development of such "over-designed" complex cranial machinery makes sense only if we assume a certain intentionality in the evolutionary process; that is, humans were accelerating their development beyond what they objectively needed for their survival at that point in time. If the human species is pushing itself (or being "pulled") forward, it could conceivably force its own self-development.

In a sense, then, natural selection is not quite so random, after all. The emerging picture of the evolution of the human species diverges drastically from the Darwinian image of species slouching toward evolvement. In this alternative schema, our species evolves not just merely to survive, but to thrive. The actors in this process, our forebears, conducted their evolution, a progressive enhancement of their cognitive and physical skills, with purpose and intentionality.

The evolution of the human species can be understood in the strict Darwinian sense, to a point. We developed and adapted in order to survive in a variety of hostile environments. Through selective breeding we gradually preselect genes that made us taller, stronger, and healthier to carry on our species' development. However, so did other species. We, though, have developed certain characteristics far superior in complexity and functionality to those found in the rest of species.

We seem to be a species in a hurry!

I label the process operating here purposive self-development. This refers to the tendency of the human species to evolve itself, through the standard natural selection process, to a level of complexity and functionality far beyond that necessary to simply survive, that is, adapt to its physical environment. We have evolved ourselves in anticipation of accomplishing a much more varied and complex set of goals, developing intellectual and cognitive abilities, especially through accelerated brain growth. These advanced cognitive abilities allow the species to develop technology, science, mathematics, and other skills, to enhance, perfect, and transform the environment and to gain further control over its own evolution.

The human species evolved into beings that had the physical and cognitive capabilities to solve more than their own adaptive challenges. They had the capacity to solve others' problems, including, as we shall see, those of the universe. Our species evolved as though it possessed a transcendent purpose and destiny.

Purposive self-development is clearly evident in our activities today. For instance, we are on the verge of discovering the entire genetic code, and soon will be able to eliminate genetic predispositions toward disease by manipulating the genes related to that disease. In the near future, we might soon be able to defeat the process of aging itself. Moreover, we may be able to program ourselves to pass on these corrected genetic patterns to offspring. As these genetic weaknesses are gradually driven out of the gene pool, we purposively advance ourselves. We are also utilizing physical manipulation to enhance our cognitive faculties. Therefore, the natural selection process is now an obviously conscious one involving research, experimentation, and conscious decision making. As mentioned in Chapter 5, we will be using such techniques to adapt to alien environments over the coming centuries. Interestingly, purposive self-development will lead to moderate modifications in the human genome at the subgroup level. Eons from now the human species may have evolved into several "offshoots" resulting from conscious changes in the body human.

Some would argue that while one can easily observe purposive self-development in man's application of twenty-first-century technology to control our evolution, it is not as easily proven that thousands of years ago *Homo erectus* intentionally evolved their brains to a state of complexity well beyond what they needed at the time. However, if natural selection holds, then we could envision members of our species gravitating quickly to and mating with those who bear the mutated genes that made them smarter, more adroit, able to develop language skills and technology. Isn't it possible that sheer *willpower* has enabled us to make choices that accelerated our development— we sense what choices we must make, and we make them? Intentionality can be deduced from available evidence (as can natural selection) over time. Moreover, it makes perfect sense if we see the human species' purpose in a much larger picture.

If intentionality is an operating principle in evolution, if somehow our species knew where it had to get to, evolution could be seen in a different light. Human evolution now becomes a vehicle by which we impel ourselves, via the process of natural selection, to get to a certain point of development. By including nonrandom intentional behavior in our concept of the process of evolution, we solidly respond to Schroeder's statistical challenge to random natural selection—we could develop ourselves into a strong, resilient, thinking being in the time allotted, say, several million years, if we were guided by, or acting in accordance with, a sense of necessity, intentionality, and direction!

The question becomes, what force could be so strong to align with and mobilize a seemingly powerful set of conditions to ensure the emergence of a species such as Homo sapiens? First physics and the laws of the universe conspire to make possible the existence of a sphere hospitable to life through a process embodied in the so-called anthropic principle. Then chemical reactions begin, which serve as a catalyst for a process that empowers life's emergence. Life begins! Humans progress at a furious, almost frantic rate, even though they could have stopped at any stage. We didn't have to become the species we are, able to colonize the solar system and rewrite the laws of physics! And we haven't stopped developing; we are accelerating our own development even to the present day. Moreover, we will strive to emerge into Kardashev's Type I, II, and III.

What induced us to relentlessly push ourselves to this stage of intelligence, of physical and intellectual development? Of what ultimate purpose were we at least vaguely aware that would require us to become what we are? Moreover, what will we further evolve into through our other adaptive mechanisms, such as species coalescence, dominionization, biogenesis, and cybergenesis? In addition, what is pushing us to attempt to vitalize the rest of the universe?

On Destiny and the Human Species

Let us gauge where we are in this argument so far. We are a species currently planning momentous events—traveling to distant stars, moving off the planet, generating Earth-like environments, and dominating nature.

In addition, the strong possibility exists that we have been in some way anticipated, planned for, awaited by the cosmos, by the universe, a position that physicists refer to as the anthropic principle. Evidence also suggests that our progression into a thinking, rational being able to achieve extremely complex tasks has been fairly rapid. Furthermore, this process was not entirely random—certainly we can witness our species making intentional choices about its own growth, a process I label purposive self-development.

Evidence suggests that the appearance of *Homo sapiens* was expected, anticipated, even desired. The remaining question is why conditions should usher onto the cosmic stage an extremely creative, intelligent being able to manipulate the environment and redirect the forces of nature? Why does life, and more specifically the human species, exist? Has the human species appeared in response to some overarching need that the universe, existence itself, had for intelligent life? Was consciousness required? Is there substantial scientifically based evidence that the human species actually possesses a purpose, a critical role to play, in the cosmos? Does the human species have a distinctive destiny?

A Universe Desperately Seeking Intelligence

Why should the universe bother to facilitate the emergence of life, and more specifically, an intelligent and efficacious form of life, the human species? What special role would our species play in the universe's existence that the cosmos would have to summon forth "the phenomenon of humankind"? The most logical answer must be that somehow the universe requires for its own survival and growth the existence of a life form, a creative intelligence.

Numerous life forms exist on the planet, many with unique physical skills and degrees of primitive intelligence. Such attributes serve them well in their efforts to maintain their survival. Yet, only one species exists with the capacity to build ships that reach the stars, to harness electricity, and construct skyscrapers. Only one exists that can significantly affect, even transform, the planet and the universe—*Homo sapiens*.

Let us look to this differentiating factor in order to begin to solve the mystery of human existence. This means that our investigations should start with a consideration of what modern science considers some of the stark realities confronting the universe. We cannot speak about the future without discussing the long-term prognosis for the state of, no, the continued existence, of the universe.

Physicists, or at least those who believe the commonly accepted dictum that the universe started as a Big Bang, have developed two or at most three possible scenarios for the universe's "end game" billions of years from now.

Some of these scenarios portend a less-than-appealing prediction for the future of the universe and any living beings within it.

In one of these scenarios, labeled the *supercritical* model, the density of the universe is such that it creates gravity strong enough to reverse the cosmic expansion. Imagine if you will this post-Big Bang expansion suddenly running out of steam. The galaxies, stars, planets, all existing matter that had been hurtling outwards since the beginning of time, suddenly rushing back, pulled by the enormous force of gravity. As this great reversal accelerates, temperatures in the universe soar, the oceans boil, the planets melt, and the stars and galaxies are compressed into a gigantic primordial atom. Physicists lovingly refer to this model as the Big Crunch.

Another scenario is equally uninviting. The universe expands forever, gradually getting colder and colder as the celestial bodies grow farther and farther apart. In this scenario, the universe eventually becomes a cosmic graveyard of dead stars and black holes, as temperatures throughout the universe drop to near absolute zero. Trillions of years into the future, even the black holes will evaporate. This scenario, called the Big Chill or entropy death, has the universe terminating its long run as a vast refrigerated wilderness.

A minority of physicists, including Freeman Dyson, envisions the possibility that a more optimistic turn of events may eventuate. In this third possibility, our universe is "precisely critical." In this universe, the expansion of the universe never reverses itself. Instead, the distant galaxies move away from us at an ever-slower pace. In this *precisely critical* universe, the neighboring galaxies are moving so slowly that their inhabitants visit and exchange material resources as well as information with their fellow *Homo sapiens* but a few light years away.

Unfortunately, most physicists see the precisely critical condition as only a remote possibility. Like Michio Kaku, they pessimistically feel that only the first two scenarios are plausible. In fact, in his 1997 book, *Visions*, Kaku doesn't even recognize the third possibility, the precisely critical universe. His two possible eventualities are the Big Crunch or the Big Chill, and "it seems as if the universe must die, and all intelligent life with it." A melancholy Michio Kaku laments on the senselessness of it all: "Such an end seems like the ultimate existential absurdity—that intelligent life struggles over millions of years to rise from the swamp and reach for the stars, only to be snuffed out when the universe dies." In Kaku's dark future, even a Type III society, which has corralled most of nature's forces for its use on a galactic level, cannot muster enough energy and resources to reverse the death of the universe.

Deriving a Mandate

While common wisdom would surmise that we cannot escape the ultimate demise of all life in the universe, that we are on a forced march to a cosmic Armageddon, perhaps we should consider an alternative perspective. If there is any force whatsoever that can muster the sheer energy and will power to overwhelm what seems to be a foregone conclusion, the demise of the universe, it is the human species. To see the feasibility of such a proposition, we must first imagine what the human species will be like, and what it will have accomplished, even a mere 10,000 years from now, let alone the millions of years in the future that these unsavory events are predicted to transpire. Even in the last century, a short time span when compared to the billions of years composing the universe's existence, we have through dominionization marshaled the energy of the atom, broken the iron grip of Earth's gravity, and harnessed electricity, and we are on the verge of controlling our own genes.

In the future, we will be able to not only elicit energy from the atom, but actually control the smallest particles that nature can offer, quarks, atoms, and protons, and commandeer them for purposes only science fiction writers can currently imagine. We already seem to be well on our way to learning the secrets of the other great force in the universe, electromagnetism. So when we consider any problem or obstacle in the future, including the unutterable horrors of the "heat death," we must factor in the serendipitous nature of human creativity, a power in and of itself in the universe. We are making nature perform in ways that it never did before—we are changing so many rules that we can only assume that we will develop methods to contravene the forces of entropy itself, in this case a Big Crunch or Big Chill eventuality.

Freeman Dyson has made many references to a time when humanity would be able to exhibit almost godlike powers to manipulate nature and the universe. In his book *Imagined Worlds*, he speaks of a distant future when "you can cooperate with your neighbors in large-scale engineering projects to keep the universe in trim and maintain the optimum conditions for life."[28] In a seminal 1979 paper, Dyson was already envisioning a universe in which humankind undermines the forces of entropy. In the worst possible scenario, the closed universe, humankind still has the capacity to overcome entropy and destruction. Dyson even provided the outlines of one method humankind could employ to save the day.

> Supposing that we discover the universe to be naturally closed and doomed to collapse, is it conceivable that by intelligent intervention,

converting matter into radiation to flow purposefully on a cosmic scale, we could break open a closed universe and change the topology of space-time so that only a part of it would collapse and another part would expand forever?[29]

The human species will consider this impending death of our universe just another challenge to be overcome. We will act, have already been acting, to devise ways to undermine the forces of entropy—even if we must totally reshape the universe itself to countermand such overwhelming negativity.

Others have expressed similar sentiments about the ever-expanding human ability to change the universe. John Gribbin and Martin Rees, in their book *Cosmic Coincidences*, have indicated that the human species may be involved in a mission, which has great significance. In their discussion of the rareness of life in the universe (they show mathematically the strong probability that life exists only on our planet), they glimpse the possibility that "living things may become an important part of the Universe."[30] This role would include "modifying the astronomical environment in the way that mankind is already modifying the terrestrial environment." If I may articulate this statement in terms of the expansionary view, dominionization, the mastery of physical and geological principles, is a necessary predecessor to the infusion of life on other planets, vitalization.

Remember, the dissipation of our universe into a state eventuating in a heat death is nothing more than the quintessential actualization of the entropy law. According to this law, systems inevitably consume energy, lose complexity, and decline to a worthless disordered condition that can never be reversed. A log can burn to ash, but ash cannot be reformed into a log. Entropy also goes by the name of the second law of thermodynamics.

Gregory Easterbrook recently stated that if entropy is a natural agent of decay, life, and especially human life, appears to be its leading foe. He quotes Stuart Kauffman of the Santa Fe Institute that Earth-borne biology represents "four billion years of defiance of the second law."[31] Easterbrook sagely remarks that while we no longer believe that humanity is at the physical center of the universe, we are beginning to awake to the fact that we are the center of meaning.

The human species, once it realizes, first, the problems confronting the universe and by definition itself, and, second, the important role the species can play in overwhelming those long-term threats to existence as we know it, will comprehend the significance of its destiny. In fact, it will come to understand its destiny as a mandate—from a moral perspective, we have no choice but to accept this mission.

We can see the slow emergence of this awareness of a cosmic destiny in our efforts to vitalize the outer spheres. As we travel from one sphere to another, even sending out our Cassinis and our Voyagers to scout the solar system and the galaxy, we are discovering incrementally how the world works. We will apply these lessons drawn from our experiences eons from now. Of course, to impact the universe in a meaningful way, we will have to occupy space and colonize the galaxies. Hence, the vitalization process is already carrying us forward in our efforts to meet our mandate.

Star Struck

Let me take this argument a step further. There are indications that the universe had an active hand in the creation of its own salvation, the development of a natural defense against its ultimate demise. Skeptical? Think it unlikely that the universe was able to summon up the agent of change that it required to challenge entropy? How could the universe manage to bring into existence "life" that would lead to an intelligence capable of rectifying the situation?

Remember that the anthropic principle suggests that just the right physical conditions existed in the universe for the emergence of life. This principle, though, does not declare that such conditions actually brought life into existence. What did, then? A likely answer may come from current theories in the field of astrophysics. Michael Meyer, Astrobiology Discipline Scientist at NASA Headquarters, offers a summary of the thinking among physicists about how life first originated on the Earth over 4 billion years ago.

NASA scientists and others have deduced that life may well be a natural consequence of stellar and planetary formation. What does it take for life as we know it to form? It requires that certain biogenic (i.e. life-forming) compounds be first created and also delivered to the proper locations where they can mix together and set the stage for life's origin. Whereas in the beginning only hydrogen and helium formed in our universe, other elements, including those considered to be the building blocks of biogenic compounds (i.e. carbon, nitrogen, phosphorus, oxygen, and sulfur) were generated in repeated cycles in which star's formed and then died. Over the course of millions of years, the process repeated itself. As a result of this star formation and death, new, fused, heavier elements were blown out into the universe.

According to this theory, only after several generations of star birth and death was there the possibility for planets to be formed with any hope of harboring the exciting carbon chemistry we call life. One of the clear indications that the chemical basis for the formation of life originated from outside this planet involves the way our chemistry operates on Earth. For years, meteorites have been shown to be a rich source of a wide variety of organic compounds that could have been critical to the origin of life on Earth. Recent analysis of the Murchison meteorite demonstrates that the way we use basic molecules, such as amino acids and sugars, could have originated in a molecular cloud, "one formed of debris strewn from the corpses of generations of dying stars." According to Meyer, such compounds crucial for the beginnings of life on this planet "could eventually be delivered to a planetary body such as Earth."

Meyer seems to suggest that the conditions of life were either directly delivered via impact on the planet's surface of meteorites, comets, asteroids, and so forth or in organic compounds "that arrived in the vicinity of Earth as it formed."

Around 3.9 billion years ago, this bombardment of meteorites and materials subsided. At this point, Earth began to form an ocean. Now, 400 million years later, we have clear evidence that microbial communities existed. Because biogenic, life-forming compounds were stable, life happened quickly and fairly easily according to chemists. In other words, as soon as Earth had an ocean, it could form life. In addition, once it had the capacity to form life, it proceeded to do so.[32]

The NASA Origins and Astrobiology programs refer to life as being a "cosmic imperative." This concept is borrowed from Christian De Duve, who in his book *Vital Dust* states that "life is an obligatory manifestation of matter, bound to arise where conditions are appropriate." He does not stipulate where these compounds and materials necessary for the generation of life originated. Nevertheless, once they do appear, life arises relatively quickly.[33]

We the Species

However, critics may respond, even if we agree that physical conditions— the universe's constants, the geological evolution of Earth, the biological conditions on Earth—all conspired to spawn *life*, how did they ensure that an intelligent, creative species such as *Homo sapiens* emerged? The answer is, although our appearance may be required, our emergence was not predestined. That, of course, is where a sense of purpose enters the picture. We pushed ourselves to develop into a being that can master the forces of nature. No one would argue against the idea that twenty-first-century humankind is involved in this process of purposive self-development. It might strain credulity, however, to claim that primitive ancestors of our species were practicing this 2 billion years ago.

If we accept the major premises of Darwinian natural selection, however, all species shape their actions in relation to survivability and adaptability. It is just that in humanity's case we defined survivability in much broader terms—we sensed, knew, recognized, that our species would eventually have to rectify problems on a much deeper level and of a much broader scope in order to survive at all. We evolved in accordance with the universe's need for some higher consciousness that would tweak and tinker with the cosmos.

Some might object that it is quite a stretch to imply that humanity could even have sensed such a mighty destiny. This is a purpose worthy of a superspecies. However, one could argue that ever since humanity, in its most primitive form, appeared millions of years ago, it has had a firm appreciation of the overwhelming power of entropy, of the tendencies of systems to run down. We knew that nature, through floods, fires, the Ice Age—could obliterate ecosystems and species. Moreover, gradually humanity has found itself in the position to ensure that systems were maintained and improved, e.g., through irrigation and rotational farming. Now, we are on the verge of controlling the weather. In two centuries, we will terraform Mars. This is merely a continuation of what we have been doing throughout our entire history, from our earliest existence up to the present moment: battling to fend off and defeat entropy before it overwhelms us.

Therefore, the concept that entropy is our enemy has always been part of our conscious development. It is also very possible that the idea of a catastrophe as all-encompassing as a post–Big Bang Big Crunch or Big Chill has been a motivating force in hurtling ourselves forward in our own evolution.

Furthermore, even if materials were not delivered to Earth to start the life process rolling, the universe did permit, even encourage, the creation of certain spheres, such as Earth, that were hospitable to ocean formation, and eventually life formation.

Some might conjecture that such reasoning raises the possibility that other beings exist in the universe. Why should life only emerge in one particular sector of one galaxy? If the need for life was so great, would not the universe attempt to create forms of life elsewhere? Even as you read this book, our species is discovering new solar systems with their own set of planets. Cannot life be brewing on these distant spheres?

Many people have determined through mathematical and other analysis that the chance for the emergence of another species even vaguely similar to

Homo sapiens is slim to none. As mentioned, Gribbin and Rees claimed that the chances that sapient life exists somewhere other than Earth are extremely low. Marshall Savage, in *The Millennial Project*, will not even entertain the possibility that life exists anywhere else in the galaxy, claiming that except for Earth, the universe is "entirely devoid of life." On the other hand, the SETI project continues to send out signals to other spheres, hoping for a response from another species. Moreover, NASA's Astrobiology and Origins projects have at least as part of their mandate the search for other life forms in the universe. The most scientific challenge to the existence of life on other planets was issued in 2000 by Ward and Brownlee in the book *Rare Earth*, which I referred to in the Prologue.

My response to the "alien life" theory devolves from my own theories on the relationship of humankind to the universe. I agree with the Russian Kadashan that as a life form advances it ought to be acquiring greater control of all of nature, including the cosmos itself. If an "advanced" civilization is out there, a Type III civilization perhaps, it has not evidenced itself to us in any tangible way—I would expect that if our cousins across the cosmos were "tweaking" and "tinkering" with the universe we might witness some of their handiwork, some sign they are creating order out of the chaos in the heavens. I see no universe building taking place in the stars—we gaze with our telescopes, but "they" do not show themselves.

Let us face cold reality. We are probably the only form of intelligent life with consciousness, regardless of how many planets exist. Some who hope to locate the wise and gentle race of superbeings who presumably will teach us how to overcome earthly problems may find such news depressing. They should take heart from the perspective Gribbin and Rees provide on this issue. To quote: "It is sometimes argued that if life is a rare accident, it would be an irrelevant fluke in a mindless and hostile cosmos. But we take the opposite viewpoint: if there is no life elsewhere, the Earth acquires universal significance as the spark with the unique potential to spread life and consciousness through the cosmos." They claim that we are only at the beginning of a process, which will take billions of years, and perhaps a literally infinite time span to run, to bring life to the Galaxy and beyond by forms of life and intelligence seeded from Earth.

I should mention that a small minority of cosmologists believes the Big Bang theory to be simply wrong, in which case corollary arguments drawn from this theory, such as the Big Chill and the Big Crunch, are also moot. The plasma physics model states that parts of the universe expand while others contract. This occurs in an ongoing pulsation that occurs when clouds of matter and antimatter collide, generate energy, and then are again repelled from one another. No cosmic explosion, no end of the universe! Another theory, the steady-state theory, claims that matter was never in that state of appreciably higher density that exploded in a Big Bang.

Therefore, the Big Bang theory has some new challengers as an explanation of how the world works. Even in these other theories' universes, though, disorder and formlessness also rule, providing us a chaotic cosmos filled with dying stars. The emerging picture of the universe shows a universe as austere and barren as it is sometime hauntingly beautiful, devoid of life and all its joys and benefits. Therefore, even in these alternative cosmologies, humanity's position as the sole source of consciousness and creativity still mandates that we endow with order and light a cosmos that offers nothing but chaos and darkness. The majestic night view of Earth from space reveals a once-dark planet now ablaze with billions of lights, man-made lights, shining like a sparkling beacon of vitality and consciousness, illumination where there was only darkness. So too will we illuminate a universe steeped in entropy, crying out for life-giving intelligence and the luminescence inherent in the human species.

Certainly, great controversy will greet any assertion that the universe "summoned" life into existence, even though evidence strongly suggests that the building blocks for life did conveniently appear. One may also fault linking the expansionary concept to orthodox Darwinian theory. In truth, there is much to explore in these issues—these questions will be debated through the decades. However, regardless of the process by which humanity first made its appearance on the cosmic stage, we now are confronted with a mandate: Preserve ourselves, and the universe itself. If the universe "consciously" tried to find a solution to its eventual chaos, then we have a beautiful story to tell. And if it didn't, the universe just got lucky! Our species is here to correct, enhance, and perfect what has been chaos, an endless cycle of star birth and death that will continue without our intervention. This mandate for action exists whether the universe made all the right moves to get us here or not.

In addition, this is the overarching reality that must condition our future actions.

However, to meet such challenges, we will have to reinvent the universe, vitalize it, and gradually create something new. We cannot merely tinker. We must infuse the universe with consciousness and intelligence. To save our universe, we must fundamentally change it!

7

The Creation of the Humaniverse

What will this new universe look like as the human species infuses the cosmos with our intelligence and consciousness? What will be its shape, contour, and character? In addition, how exactly does our species propose to meet the technological challenges necessary to achieve its golden destiny?

Generally, we vitalize the universe by extending our humanness first across our magnificent planet Earth, and then to other planets, throughout the solar system, and eventually to other star systems. We will create man-made orbs, artificial superspheres, to fill the open spaces in the universe, and terraform the planets. We will build cities and empires that span the galaxies, develop a multitude of new cultures, and invent new technologies. Meanwhile, the human species will multiply, filling the universe with billions if not trillions of human beings. This multitude of individuals will illuminate the spheres and animate the cosmos. And as we enhance the universe we will transform ourselves.

The will and spirit of mankind will create a novel entity I label the *Humaniverse*, an interconnected entity infused with human consciousness and permeated by rational creative activity. The appellation *Humaniverse* captures the essence of an entity whose very existence is changing in the direction of a species extending its physical and intellectual selves into the cosmos.

In truth, we are already creating the Humaniverse. Through such processes as *biogenesis, cybergenesis, species coalescence,* and *dominionization,* we have rationalized and enhanced the Earth and its surrounding stratosphere and ionosphere. Furthermore, we will soon begin in earnest vitalizing the solar system during lifetimes of many who are alive today. The terraformation of Mars and other orbs will be a key component of this enterprise.

Our current efforts to create and invent as chronicled throughout *The Future Factor* indicate that the human species already possesses both an intuitive sense of its ultimate purpose and a willingness to embark on this great mission. Now we must devise a fully articulated schema of our destiny and how to achieve it. Even now, forces are gathering throughout our world to ensure that the issues around human destiny and its fulfillment gain a full public airing. They are designing transgalactic human settlements, and drawing up blueprints for rockets that can move at half the speed of light. Groups with names such as the Mars Society, the Extropians, the National Space Society, the First Space Foundation, and the Planetary Society are trying to educate the public, venture capitalists, and legislative bodies about the nature of our destiny and seize the opportunity to realize it. They work tirelessly to turn this dream of a humanized universe into a reality we can all participate in.[1]

Let us examine more closely the new world they are ceaselessly working to create.

Visions of the Humaniverse

We will be extending our physical and intellectual selves to planets, spheres, galaxies, and star systems for many eons to come. Before I describe the shape and character of this humanized universe, let me address some of the scientific and technological issues confronting our nascent spacefaring society.

The two principal elements that we must contend with are space and time, the space between our planet and other destinations, and the time it takes to reach these distant worlds. In spite of humankind's expanding technological wizardry in fields such as aerospace and astrophysics, we are still overwhelmed by the distances between bodies and intimidated by the sheer size of the universe. An old adage asserts that the problem with the universe is not that it is larger than we think, it is larger than we can ever imagine. As we peer into the cosmos through the lenses of the Hubble Space Telescope, we see before us a massive expanse of star clusters and galaxies that seems to go on forever. Moreover, we suspect that these instruments reveal only a fragment of the whole picture. We might wonder whether this entity contains any outer limit.

How big is the universe? To measure cosmic distances, we commonly employ a standard called the astronomical unit (AU), the distance of the Earth to the sun, roughly 93 million miles. Mars, Venus, and Mercury are far less than one AU from Earth, and Jupiter and Saturn about 10 AUs from Earth. The very edge of the solar system is 100 AUs from our home planet, about 9 billion miles away. We measure distances also in terms of the light-year, the speed at which a light wave travels in one Earth year. The speed is so enormous that it is usually stated in miles per *second*, not years—light travels at the speed of 186,000 miles per second. To give you a better appreciation of this speed, let me translate this to the light-year. In one Earth year, light travels almost 6 trillion miles.[2]

To become an intergalactic civilization, we must be able to travel at velocities approaching the speed of light. Surprisingly, in the next few decades we might design crafts that can reasonably approximate such speeds. If we do, we could quickly colonize the solar system and begin to evolve the Humaniverse.

Such distances may seem daunting given current spacecraft technology. While our contemporary chemical engines can safely get us to the moon and Mars and return us securely to Earth, they would have a hard time transporting us to the outer reaches of the solar system, let alone beyond, in a reasonable period. While our Cassini Project will indeed reach Jupiter and Saturn, the craft depends on various nearby planets' gravitational pull to propel it on its longer trek into the cosmos. The journey of Voyager I, launched September 5, 1977, illustrates the difficulty of traversing great distances. Voyager, traveling at 51,000 mph by 2000 was only 10 light *hours* away from Earth. The closest star to Earth is Proxima Centauri, 4.3 light-years or 25 trillion miles away from Earth. At its current speed, Voyager would take 74,000 years to get to Proxima Centauri![3]

Realistically, humans must be able to reach destinations in space within a reasonable period—two, three, perhaps ten years at most. After that point, a variety of problems arise around issues of fuel, food supplies, not to mention the psychological difficulties experienced by astronauts separated from Earth-bound civilization for protracted periods. Fortunately, new technologies may solve the mechanical obstacles preventing human species from traveling to distant planets. We can be reasonably sure that soon we will construct spacecraft exponentially faster than Voyager or any other craft we have constructed.

In May 1999, Homer Hickam, Jr., threw down the gauntlet to the space agencies and the public to embrace a new generation of space rocketry that will enable humankind to colonize the cosmos. He had recently visited NASA headquarters in Huntsville, Alabama, "Rocket City, USA," and observed on their computer screens the rocketry of the future, craft that would make chemical rockets seem absurdly primitive. As Hickam exclaimed, he "came away impressed not because these systems are so exotic or futuristic but because we are quite capable of building them right away." He studied

NASA's blueprints for rockets capable of reaching velocities of between one-third and one-half the speed of light. He wrote in *The Wall Street Journal* that NASA must begin to lay out a 10-year program to produce a working advanced propulsion space drive based on these new technologies. Hickam, a retired NASA engineer whose early interest in rocketry was chronicled in the inspirational 1998 film *October Sky*, still retains the spirit of destiny that originally drove him into the field of aerospace. He proposes that "we take it upon ourselves to move out into the solar system and conquer, settle, and prepare it for all the world to follow."[4]

Homer Hickam has reason to be encouraged. According to scientists at NASA's Jet Propulsion Laboratory, we will someday find it relatively easy to visit the stars. They claim that we already possess the technology to develop interstellar launches that could travel one-third the speed of light or even faster. At this velocity, a craft could cover an incredible 2 trillion miles per year. Dan Goldin, NASA Administrator, thinks we will be able to launch interstellar missions in the next 25 years.

One type of craft that Hickam saw in NASA's computer is the *lightsail.* Forward Unlimited, a Clinton, Washington, company is working on such a concept, which they label the *solar sail.* Picture a craft, a rocket, surrounded by a huge sail, possibly 6–10 miles wide, yet only a few atoms thick. The craft is propelled by particles of light called photons that hit the solar sail and push the spacecraft forward at incredible speeds. Lasers that orbit the sun are transmitting the photons. These orbiting lasers, which draw their power from the sun itself, can continue to push the sail spacecraft forward at incredible speeds for months or even years over millions of miles. Theoretically, this laser-powered solar sail craft could travel at one-third the speed of light, making the trip to a star system 4.5 light-years from Earth in about 20 years. In that time, our craft would have traveled an almost inconceivable 20+ trillion miles.

More astoundingly, these sun-orbiting lasers, generating power in the gigawatt range, would simultaneously perform other functions. The orbiting lasers will be removing Earth-orbiting space junk, providing energy for our planet, and maybe even destroying Earth-threatening comets and asteroids. According to Henry Harris, a JPL physicist, "The ultimate instrument for interstellar travel will not be merely for travel."[5]

Another suggestion for achieving velocities near the speed of light is an *antimatter*-based craft. When antimatter and matter react, their impact results in the creation of the highest density of energy known to man. The antimatter-matter reaction could release charged particles out the back of a spacecraft for thrust. These charged particles move very fast, approximately one-third the speed of light. To be able to successfully use antimatter-based energy for space travel, we must overcome daunting technological obstacles. At this point the scientific expertise does not exist to create a nozzle, the back end of the rocket, that can withstand the intense heat such an antimatter-based engine would be shooting through it. Moreover, a pure antimatter engine would require thousands of tons of antimatter, plus matter, on the order of the size of the Washington Monument. We would probably need to develop a manufacturing infrastructure in space to harvest the antimatter.

Developing such a new form of energy would require a major investment of time and money. NASA and space enthusiasts realize that it is easier to sell such a project to the public and the government if they can show how it could have multiple applications. In this case, the infrastructure we will use to produce antimatter would have an immediate application, a medical one—antimatter is known to image and destroy certain cancer tumors. Even members of the public and government officials lukewarm to the idea of space exploration would support new technologies guaranteed to eliminate one of the dread diseases plaguing our species.

Nuclear fusion reactors have also been suggested as the way to send our rockets into the cosmos at nearly the speed of light. Antimatter could be used to ignite such a fusion reactor. Some technical problems must be overcome in applying fusion to rocket propulsion. In addition, as we chronicled earlier, our civilization has not yet perfected the basic technology of the nuclear fusion reaction itself.[6]

The development of ultra-high-speed engines will make interplanetary travel accessible to all. Within the century we may see hosts of expeditions to the farthest reaches of the solar system as well as to the asteroid belt between Jupiter and Mars. While governments and corporations will underwrite such expeditions, private groups, families, and even local churches who pool their resources may fund launches in an effort to stake a claim on a nearby asteroid.

Breakthroughs in the technology of speed make rapid development of the Humaniverse a high probability. Our species will take two separate pathways to develop the Humaniverse. We will initially explore and colonize spheres from which we can return to Earth in a reasonable amount of time. For instance, those who colonize Mars could navigate between Mars and Earth in a few hours' time in their ultra-high-speed spacecraft. An antimatter-based spacecraft will enable the Earth-bound local commuter to travel to the edge of our solar system, about 9 billion miles away, in a few days' time. In this first scenario, even as we colonize other spheres, we are still staying relatively close to home.

However, eventually we will be colonizing planets and star systems so far from Earth that we will return to the home planet rarely if ever. As these colonists' descendants migrate to even more distant spheres, perhaps in other galaxies, a return to Earth in years or even centuries might be impossible. Over the next millennia, humans will gradually migrate from one sector of the universe to the next, following the historic patterns of human migration on Earth. Eventually, such settlers of the universe will no longer refer to the Earth as a "home planet"; in fact, they might not refer to Earth at all.

As we observed in the last chapter, many thinkers and futurists have projected the human future in the cosmos, usually with timetables and the like. Marshall Savage, in his book *The Millennial Project*, presents one of the more detailed maps for the development of a Humaniverse-style future. As we saw earlier, Savage the space activist works diligently to generate enthusiasm and funding for space colonization. His First Space Foundation actively engages in research, fund-raising, and education to further the cause of the human colonization of space. In his book, Savage presents realistic and convincing scenarios that describe the settlement of the cosmos.

His vision of the future is compelling in its enthusiasm and overwhelming in its detail. Combining New-Age dreaming and hard science, Marshall Savage weaves a fairly elaborate account of the next several centuries of human migration into space. His stepwise plan for the colonization of space makes sense. He suggests that humans first practice on Earth the basics of space colonization, preferably by building islands in tropical oceans. One of his proposals has us "growing" floating sea colonies in the middle of the ocean. Because they acquire their food and energy from the sea these colonies will become self-sustaining. Actually, these sea colonies are more living organism than fabricated structure.

In Savage's universe, the species would construct a bridge to space and "ride a rainbow hued array of lasers into orbit." Actually, his vision could be realized if the solar sail ever becomes a prototype for all future starships. He thinks we will then establish habitable ecospheres in space, similar to the O'Neill space modules. The next step is the terraformation of Mars, making of that desert world a vibrant living planet. His larger step, which he labels "Solaria," involves human migration throughout this solar system, transmuting raw materials into living tissues. The last stage is "Galactia," which involves the remote seeding of planets, star travel, and other intriguing activities.[7]

To give an idea of how expansive the Humaniverse might be, let us look at some of Savage's population projections over the next thousand years. Curiously, Savage is one of the few futurists to actually attempt to project the size of the transgalactic population a thousand years hence. Using some basic mathematical extrapolation of an historic growth rate of 2 percent per annum, Savage predicts that 1500 years from now 100,000 billion people will populate the solar system alone. His world has relatively small ecospheres—asteroids, planets, and artificial islands in the middle of space—featuring populations of various sizes but averaging only 100,000 people. (Message: We realistically cannot expect to colonize the solar system, or even a planet the size of Mars, unless we dramatically increase our population numbers beyond the woefully inadequate 6 billion people living on planet Earth.)[8]

These billions, even trillions, of humans living in the solar system will reside on terraformed planets, asteroids encased in membranes, artificial biospheres circling planets and the sun. Seen from afar, the solar system will look brighter, more interesting, fuller, than the mostly darkened void that greets our telescopic gaze today. It will be a rich world, a manifestation of the best qualities of the human species. Most importantly, it will be permeated with humanity.

As we venture past the Kuiper Belt, 1000 times farther than the Earth is to the sun, and migrate to other star systems, we will leave traces of our species in the form of colonies and outposts. The universe will slowly evolve into the Humaniverse, acquiring the human traits of rationality, consciousness, and beauty.

The Gift of Consciousness

As we migrate from planet to planet, star system to star system, we build cities, reshape the planet, defining these spheres geography, topography, and climate. We will illuminate these spheres as we have brought light to Earth, and ignite the skies with our presence.

The human being is uniquely equipped to create this superorganized *ultra-*intelligent structure I label the Humaniverse. I say this not because we are bigger and stronger than other beings, or quicker, faster, or more physically adroit. It is our higher intelligence and higher consciousness that empowers us to dream about reaching the stars, changing the cosmos, and actually developing the technology and science to make these images become reality. In addition, it is our consciousness with which we are infusing the universe.

Let me make clear what I mean by this property called *human consciousness*. Consciousness is the human's ability to think in terms of both complexity and abstraction. The term also refers to our unique gift to approach problems creatively, and the ability to mentally project ourselves out of our own environment and cogitate on the conditions in another country, planet, or star system. The human being also has the ability to think and act on multilevels of reasoning and reality, and to approach problems from different angles. Most importantly, it is the property that enables us to "know that we know." (In the next chapter I describe how this differs from various states of awareness claimed for other entities such as animal and machines.)

An important component of human consciousness is what we refer to as "intuition." Many observers describe this quality as a form of "super reasoning," which enables humans to draw conclusions about situations and solve problems with only a bare minimum of facts and a short time to analyze those facts.

Psychologist Gary Klein has intensively studied intuition, which some consider a sixth sense, a form of superconsciousness. In his book *Sources of Power: How People Make Decisions*, Klein describes the decision-making techniques Army pilots, firefighters, and air traffic controllers use in stressful situations. Klein discovered that while these people usually made the right decision in high-stress situations, they later had difficulty articulating to researchers how they actually had arrived at such decisions. Klein discovered that firefighters, in perilous rescue situations, almost instantaneously could reach into past personal experiences and race through memories of former situations for a match with the problem of the moment. They did this so rapidly that they could not see or experience themselves doing it. Invariably, they would make accurate risk-assessments about whether to race into a burning building and determine strategies for maximizing success. Klein discovered that neonatal nurses similarly maintained intricate inventories of cues and phenomena that signaled the need for action during emergencies. This intuitive intelligence is so interwoven into human activity that it is difficult to deconstruct, mechanize, or reproduce it artificially.[9]

Human beings' talent for thinking intuitively makes them well-suited for colonizing space. When the human species explores the cosmos, they will be greeted with the unknown, the quirky, and the uncanny. Our ability to sense the right direction and the right solution, as well as our capacity to process data and stimuli on multilevels, will enable us to survive in and ultimately change and enhance a universe that operates according to the unpredictable and counterrational quantum principles. We have learned throughout our history to not only expect the unexpected, but to overcome and master it. In fact, our natural curiosity conditions us to embrace the serendipitous.

Human consciousness and intelligence will enable us to reach the stars. At the same time, this very consciousness is our gift to the cosmos. Imagine a dead universe suddenly contacted by an intelligence which has over time learned how to manipulate matter and corral the basic secrets of energy. Now picture this sphere being occupied, organized, rationalized, embellished, and made beautiful by human intelligence. We are effectively extending and expanding the noosphere to the farthest reaches of the solar system—human creativity, spontaneity, and serendipity will change the very appearance of the system, animating planets and vitalizing the void.

As we migrate through the cosmos, building our cities, enriching the landscapes of planets and star systems with forests and oceans, we too will grow and evolve. The challenges we must confront and overcome will serve to expand our knowledge, skills, and our very creativity. The very process of becoming "masters of the universe" serves to amplify our intelligence and magnify our abilities. We will grow as individuals and as a species.

We will also evolve as we come to gain more control over *time* and *space*. A thousand years from now our efforts in the area of biogenesis may have humanity at the doorstep of immortality. Imagine how our perspective changes regarding our lives, our goals, and our families. We will attain a concept of control over nature and over ourselves that we living at the dawn of the third millennium can hardly fathom—we will control our cells, our genes, our atoms.

Our enhanced control over space will profoundly enhance the view we have of the very meaning of being human. The average person will be able to effortlessly traverse trillions of miles in a relatively short time. This ability to glide casually over enormous sections of the cosmos will tempt us to experience the sense of being almost godlike. Over time as we move near the speed of light we will not feel that we are traveling *to* the universe; we will perceive that we are walking the grounds of our Humaniverse. Our ability to conceptualize in terms of trillions of miles will permanently and profoundly alter our view of the relationship of humanity to the cosmos—over time we will unconsciously achieve a sense of ownership of the territory we can physically navigate.

As we evolve intellectually and physically, our power to control gravity and electromagnetism will expand immensely. We will become by definition something different, something more sublime. The division between what we have traditionally considered the "divine," the transcendent, and the human, will lessen, even perhaps become irrelevant. After all, a species that has mastered the fundamental dynamics of the universe has evolved beyond the merely human.

Species Coalescence, across the Cosmos, over Time

As we evolve, our consciousness will become more sophisticated, more complex. The Humaniverse, human imagination and ingenuity, will become increasingly complex also. It is, after all, a reflection of our own imagination, wishes, and desires.

One of the problems that will inevitably emerge is an ironic outgrowth of our own success at adapting the universe to our needs. Initially, our space colonies will be situated relatively close to each other. Therefore, the early settlers will be able to communicate with inhabitants of other colonies via radio and other technologies and visit other colonies quite easily. From the perspective of travel and communication, each settlement will be no more distant from other colonies than a citizen of Canada is from an inhabitant of the United States.

However, over time, over thousands of years, maybe even sooner, humans will develop the enhanced technological capacity to migrate into distant quadrants of the universe. Yes, there will be trillions of us; but each colony may only be able to communicate with or visit a small percentage of our fellow humans. Radio communication is of limited use—the waves travel too slowly to make communication between civilizations on different star systems feasible. Will our civilizations continue to stay in touch? Will our future civilizations even be aware of each other several centuries from now?

This inability to communicate and interact will have profound implications for human development and the continuity of the human species. Imagine a time when enclaves of humans, even colonies of millions, are living light-years from each other. Over the centuries, these colonies would diverge into different cultures, developing their own languages, science, clothing, literature, and art. This diversification along cultural lines does not necessarily endanger the process of human continuity—on Earth peoples of different cultures and traditions still recognize the others as inherently human. Further contravening human solidarity will be the tendency for the distant species to gradually diverge physically. Different colonies and civilizations living in different climates and galactic environmental conditions may intentionally evolve themselves to adapt to these new conditions. Alternatively, over time the physical shells of the humans in each colony might naturally evolve in response to various environmental factors on their planets or in their solar systems.

Suppose the changes in each colony are such that the descendants of each colony become so different physically and culturally that it would be difficult to label as *human* the resulting totality of subspecies and offshoots. If these descendents of humanity should someday meet, would they fail to recognize members of the other colony as fellow humans? Will we have diverged to such an extreme that we no longer consider each other family members, not even distant cousins?

If this happens, real problems would exist for the further development of the Humaniverse. After all, could the future evolved interplanetary system still be considered a Humaniverse if it is being developed by species that no longer share the same species consciousness, who no longer can be considered in a state of coalescence with other members of the species? If each of these variant species has become that different, can they all be considered human?

My answer is a definitive yes! I believe that as members of our species move so far apart that they can no longer interact with each other physically, verbally, by space flight or radio transmission, another mechanism will kick in which will ensure that the human species maintains its continuity. We will do so regardless of how much we now differ physically and/or culturally. I believe this mechanism will help the human species maintain a common identity long after we migrate to remote galaxies and star systems.

This mechanism to which I refer, *nonlocality*, is only vaguely understood at present, and is the source of considerable scientific controversy. This concept has emerged from studies of quantum physics which demonstrate that nonlocality is a basic operating principle in the universe.

The term *nonlocality* first emerged in physics in studies of the behavior of tiny particles. One of the strangest claims of quantum mechanics is that two particles can be so entangled—inextricably bonded—at birth that their behavior will forever be linked regardless of where these particles drift. In theory, a measurement on one entangled particle is linked to a measurement on the other, even if each particle may have traveled to opposite sides of the cosmos.[10] While such a finding defies common sense, empirical research strongly supports such a conclusion.

According to the late Irish physicist John S. Bell, if two subatomic particles once in contact are separated to some arbitrary distance, a change in one is correlated with a change in the other—instantly and to the same degree. No "travel time" exists for any known form of energy to flow between them. That is, they don't have the time to signal to each other "I have changed, so why don't you change in the same direction?" Yet experiments have shown these particles do change in the same direction, and do so instantaneously. Neither can these nonlocal effects be blocked or shielded—one of the trademarks of nonlocality. Now, physicists at the University of Innsbruck in Austria have created the same eerie link among a trio of photons, so strong that behavior of the first two photons preordains the result of the third measurement.[11]

What does the behavior of photons have to do with people? Everything, according to many psychologists and physicists! The concept of nonlocality will help us solve the conundrum confronting our nomadic humans of the far future attempting to maintain species coalescence.

Studies of the behavior and personalities of *twins* are leading to some fascinating conclusions about human consciousness. The University of Minnesota's Center for Twin and Adoption Research, founded by Dr. Thomas J. Bouchard, has studied separated twins throughout the world. Over 100 pairs of twins, all separated at birth and weaned and nurtured by different sets of parents, have been studied at the center by psychologists, psychiatrists, cardiologists, dentists, ophthalmologists, pathologists, and geneticists. These professionals have examined every aspect and trait of these twin pairs, including their blood pressure and dental histories.

Researchers at the Center repeatedly discover that twins often develop many of the same physical and personality characteristics, such as heart murmurs, allergies, and IQs. Such similarities are usually attributed to the fact that each twin in a given pair shares much of the same genetic material.

However, while genetics might explain many of these behavioral and physical similarities, it cannot serve as an explanation of a deeper mystery related to these twins: studies repeatedly show that the two members of these separated pairs, who have never met (outside of the womb, that is), exhibit uncanny similarities in lifestyle. For instance, twins have been found to use the same brand of shaving lotion and tooth paste, and smoke the same cigarettes! In one twin pair, each member was shocked to discover that the other twin stored rubber hands on his wrist, entered the ocean backwards and only up to his knees, and struck the same type of pose for photos.

Some as yet unknown force seems to be influencing such twins to grow up "together," even though they live physically apart. Many of the more adventurous experts are looking for an explanation of this phenomenon in the aforementioned findings in quantum physics. They reason that perhaps such twins are replicating the behavior of subatomic particles who themselves act similar to each other even though they have been separated at birth. The idea of applying the activities of the subatomic world to human behavior was proposed not long ago by Nobel physicist Brian Josephson of Cambridge University's Cavendish Laboratory. Josephson contends that nonlocal events at the subatomic level—for example, the fact that subatomic particles, even after they are separated, maintain the same rate of spin—may emerge in our everyday experience.

Some physicists have used the term *nonlocal mind* to account for some of the ways consciousness operates. Some have suggested that this nonlocal mind would not be completely confined or localized to specific points in space or time, but exists over a broad range of geography, forming what Nobel physicist Erwin Shrodinger calls a "unified field." If human minds act as subatomic photons do, each performing the same as the other, "sensing" the behavior of the other regardless of the physical distance between them, then we can begin to fathom the mysterious synchronous behavior of separated twins. (Some physicists even think this nonlocality principle can explain phenomena we have heretofore described as "telepathy." That is, when person A exhibits telepathic behavior, he is not so much "reading" person B's mind; rather, A's mind is merely tuned into a unified field that both he and B are part of or participate in.)[12]

The theory of nonlocality has profound implications for our descendants thousands of years hence, living in colonies so far apart that large segments of the species can no longer communicate or have contact with each other. It is safe to say that many of these colonies will exist only as myths in the other civilizations' fables and prehistories. Nevertheless, all these colonies will still be growing, maturing, and evolving along similar patterns, as our twin pairs have. Humans no longer living in each other's galaxies will still be co-developing their common Humaniverse, sensing, even experiencing, each other nonlocally, replicating the "subatomic spin" of the others' civilizations as we move together to carve order out of chaos and illuminate the universe.

Equally importantly, because each of these subgroups of our species, geographically distant, still maintains their humanness, they in fact will equally contribute to the creation and expansion of the Humaniverse. The qualities they are extending throughout the universe will essentially be the same qualities, human consciousness and intelligence, that the other species are spreading across the cosmos.

Even 100,000 years hence, a human living on an asteroid or a moon of Jupiter, encountering another being descended thousands of years earlier from the human species, will recognize the inherent *humanness* of that being, regardless of their differences in culture and physiognomy. In fact, one could make the point that as we grow apart, our sense of humanness may actually increase as we are forced to forge an identity that retains the greatest property of all, our humanness.

The Humaniverse is both physical, an entity composed of connected planets and space islands, and mental/spiritual—a sensing connectedness that transcends the physical. Of course, if one branch of humanity is adapting to conditions unique to its specific galaxy or planet, the other branches need not follow—each will pick and choose and similarly discard unneeded modifications. Nevertheless, we will still be growing together as a species.

The Humaniverse, Not Gaia

The emergence of human consciousness and human intelligence is a unique historical event—the human race's capacity to vitalize, bring life, order, creativity, and novelty to everything it touches, sets the world on a completely new evolutionary trajectory. Moreover, the world now possesses an entity, the human species, that could develop the tools to save the universe from the Big Chill or the Big Crunch, the demise augured by the Big Bang theory.

Hence, *human will* is the ultimate determinant of the shape and direction of the universe. It will not be left to the universe to determine its ultimate fate—it has no concept of where it is going. At best, it will settle into a moribund chaos, at worst it will teeter on the edge of dissolution and destruction. The human being has a different destiny in mind for the cosmos— we are actively engaged in creating a Humaniverse of our own.

This state of affairs leads us to repudiate the common late-twentieth-century bromide that the human species is living "in balance with nature." Nature has already given us indications of its possible overall directions, usually a variation on entropy. One would imagine, then, that those championing an aggressive program of human space exploration and settlement along with the transformation of the universe would not invoke visions of the cosmos that encompass a passive or conciliatory approach to nature. Sad to say, some enthusiasts for human space colonization harbor sympathy for just such a philosophy, one that has the potential to constrict our species' advancement and imperil the fulfillment of human destiny.

This philosophy goes under a purportedly scientific maxim known as the *Gaia Principle*. Dyson, Savage, and others make constant reference to this principle when discussing the weightier issues of man and the universe. The principle is based on the idea that many systems, including Earth, are a collection of delicately balanced interdependent systems. The Earth is made up of oceans, atmosphere, radiation of the sun, and the activities of living forms such as man, each complexly interrelated. A drop in global temperature can cause significant amounts of the ocean to be added to the polar ice caps. Volcanic activity pushes more dust into the skies, which would lead to the formation of tiny droplets of sulfuric acid in the upper atmosphere. The Earth may cool as a result.

In the mid-eighteenth century, Scottish geologist James Hutton decided that enough interdependence existed to support the theory that the Earth behaved as a *single living organism*. James Lovelock, in conjunction with Lynn Margulis, turned this contention into a full-blown cosmology, which gained many adherents. This fashionable cosmology sees the Earth, as well as every galaxy, solar systems, and ultimately each universe as a single living organism (Gaia), replete with mechanisms of self-defense.

This fanciful concept, while lovely in its expostulation, holds little water with scientists. As Richard Dawkins and others have pointed out, the Earth can hardly be considered an organism. For one thing, it does not reproduce. It may be a system, yes, but it is not an organism. Brian Silver, in his *Ascent of Science*, asks whether we gain anything from defining the Earth as a single organism, since we already understand the interplay of the different phenomena on Earth, such a climate and volcanoes. He vociferously objects to one point in the Gaia hypothesis, that the Earth "defends itself" as an organism does. Silver says that there is little reason to use such metaphors to describe a fairly predictable behavioral pattern in which the Earth and its inhabitants adjust to external and internal changes, in many cases successfully.

Part of what attracts adherents to this theory is the imagery with which its proponents embellish it, the aura of Earth and universe under the influence and protection of a goddess, Gaia. New-Age proponents have taken Gaia to their bosom. Silver sarcastically asks whether New Agers would have been as attracted to this theory if instead of being named after a Greek goddess, the theory had been labeled "coordinated interaction of nonlinear dynamics in the terrestrial bio- and geospheres."[13] I personally doubt it.

Marshall Savage refers to the Gaia concept throughout his book. Marshall uses Gaia-like terminology in his discussion of the colonization of Mars. "In the anatomy of a living planet, the rocks are its bones and the atmosphere is its skin. Mars still has a bare skiff of skin left on him."[14] Portraying the planet Mars as a body necessarily leads Savage to envision humanity as a healing element—we will add life to make the patient better. Dyson, in *Imagined Worlds*, warns humankind that "we have to learn the art of living on Gaia's time-scale as well as our own."[15] He speaks of a time a million years hence when humankind ceases to exist. He trumps this questionable prediction with the statement that "we shall still survive in equilibrium with Gaia if we survive at all." In other words, life, or Gaia, will go on, even if we don't survive.

Before we dismiss the Gaia concept as merely a fanciful leap of the imagination, harmless in its effects, let us remember that some of the world's most powerful individuals and organizations strongly endorse and actively support the Gaia concept. The United Nations has endorsed the Gaia concept in everything but name. At the 1992 Earth Summit, the UN signaled its shift in emphasis from the pro-development philosophy to one based on environmentalism and sustainable growth. Read the two documents emanating from this summit, *Agenda 21* and the *Global Biodiversity Assessment*, and one can detect the fingerprints of the Gaia movement on every page.

Charles, the Prince of Wales, was an early convert to James Lovelock's Gaia principle. According to researcher Joan Veon, in her book *Prince Charles: The Sustainable Prince*, Charles early in his life became an adherent of Lovelock's concept of "holism," the "principle of balance, harmony, and the interconnectedness of nature with the search for inner awareness." According to Veon, Prince Charles's conversion to the Gaian view of humans as merely an equal to the earth, plants, and animals, set the stage for his lifetime of involvement in the promotion of alternative medicines and organic farming. Charles, known by many as "the Green Prince" and the "Eco-Prince," is now immersed in and supports activities across the globe aimed at bringing national and international policy into harmony with the Gaian principle. Veon chronicles his highly effective efforts to gain support among governments, business leaders, academics, and key media figures for the Gaian-inspired principle "sustainable development."[16]

Moreover, many of the people involved in envisioning and implementing our spacefaring future, and currently in control of the purse strings of space projects, give credence to Gaia as a concept and an operating principle. As NASA and other groups evolve their Origins and Astrobiology programs, Gaia could be a guiding *zeitgeist* informing their decisions. (Astrobiology program engineers use the term Gaia in their literature.) If this concept becomes a philosophical underpinning of space travel, and we adhere to the spirit of Gaia, we will never create the Humaniverse. The reasons are many.

First, the Gaia hypothesis envisions man as part of nature—the Earth is a living organism, and therefore humanity is a part of this living organism. However, Gaia misses the point completely. Although we evolved from nature, our consciousness and intelligence, our ability to create scientific and artistic outputs, demonstrate that we stand outside of nature even while physically a part of it. We are wholly unique and new in the universe's evolution, unparalleled in terms of intelligence and creativity. Unlike the Gaians, most of humanity has understood that whereas nature is at times an entity we can work with and through, often it is something we must transcend. This century, we even overwhelmed gravity, one of Gaia's pet forces, when we decided it was time for humankind to visit and explore other spheres. Moreover, in complete violation of the Gaia Principle, to build the Humaniverse we will require and induce nature to work at our speed, not us at hers.

Second, the Gaia hypothesis puts the existence and maintenance of the system ahead of the health and well-being of the system's inhabitants. This explains the almost religious fervor with which ecologists and environmentalists attack human growth and progress—they see human advancement as the ultimate threat to Gaia's integrity. Gaia's acolytes will take whatever actions necessary to preserve Gaia, even if such action causes human discomfort and misery. In the Gaian hierarchy, the needs of even the lowliest creatures and the smallest brook will be given precedence over the needs of humanity.

A true adherent of the Gaia philosophy would regard as sacrilege the tendency of mankind to intrinsically change, tweak, readjust, and even radically transform elements of the Earth and soon the outer spheres. From their perspective, we are tampering with the integrity of a balanced organism, the Earth. Frankly, if we use Gaia as a guiding ethos, we will never change the universe—by definition, it is all, in a certain chaotic sense, in balance. In the mind of the Gaia true believer, mining the Earth, splitting the atom, and splicing the gene are all forms of tampering. Imagine the fire and brimstone that Gaia will rain on the heads of humans when we master nanotechnology, a science that has us taming individual atoms and recombining them into food or manufactured goods. Moreover, those who endorse the Gaia principle in toto cannot logically support spacefaring and star travel, processes that by their very nature require humankind to tinker and tamper to make the universe its home.

In a perverse way, adherents of the Gaia ethic use this principle to excuse, or at least explain, horrors such as poverty, mass starvation, and AIDS. Although Marshall Savage waxes eloquently about the impact of man on the cosmos in the distant future, he portrays our current actions on Earth in less charitable terms. He accuses humanity of ravenously consuming the last vestiges of vanishing natural resources, "spewing out mountain of garbage and rivers of toxin in exchange." The result, he says, is a dying planet. If we don't mend our ways, this living organism will do what any other organism would do—it will defend itself. "From Gaia's perspective, the answer to this disaster is a species-specific plague to wipe us out—AIDS perhaps, or maybe something even worse."[17]

The belief that God or nature punishes the so-called excesses of human behavior with lightening bolts, volcanic eruptions, or AIDS is an old one. The Gaian hypothesis is just its New-Age permutation. Yes, diseases do ravage humanity from time to time. However, we have no proof that this exemplifies the "living organism" Earth acting to defend itself by reducing population size. If this were the case, would we not be duty bound to let nature have its way with us? In addition, are we not thereby frustrating the wishes of Gaia by refusing to obediently die in droves? Is humankind sinning against Gaia when it discovers a cure for AIDS? We thwart potential starvation of our ever-growing population by bioengineering better crops. Is it more moral to go easy on agricultural development and let Gaia do what she will with our "excess population"?

We will only achieve our destiny when we adopt the mindset that humankind, not Gaia, sets the rules for the future. The Gaian philosophy encourages an approach to nature that borders on the obsequious. It kills in its sleep our propensity to risk, warning as it does that any human excess will incur Gaia's wrath in the form of plagues and pestilence. If we are to vitalize the cosmos, we must allow our human imagination, the novel additive to the operation of the Earth and the universe, to reign supreme.

Third, the Gaia hypothesis tends to see all systems—galaxies, planets, and solar systems—as "living." In the process, the Gaian notion blurs the key difference between Earth and almost every other entity. Earth, and Earth alone, is filled with life, air, microbes, trees, water, and most importantly the human species. Proponents of the Gaia Principle see the mythic Gaia operating everywhere, on Mars, asteroids, the sun, effectively denigrating all that makes the Earth special. Worse, by dumbing down the notion of life, proponents of Gaia eliminate the primary need for the human species to visit these spheres. After all, if as Gaia states, these entities, such as planets and galaxies, are already integrated "living" organisms, what does humanity have to offer? Why should humanity even expend the energy to go there?

We should go there precisely because these entities are dead, chaotic, lifeless, and awaiting the majestic touch of humanity. Let us admit that while Mars has form and substance, it does not, as Gaia theorists would have us believe, possess life as we know it. If the red planet hopes to achieve such a state, if it hopes to be vitalized, infused with life, have its oceans revived and its terrain enriched, it must await the arrival of humanity. Moreover, the universe, while filled with movement, bursts of energy, cosmic collisions, and exploding stars, must stand vigil until a sapient species endows it with consciousness.

Moreover, when we go to these outer spheres, we are not doing so to achieve what Gaia calls the "greening of the universe." We are not traversing the galaxies in order to serve as nature's cosmic agent. We extend these efforts in order to bring to the cosmos a quality that never existed in the universe before humanity appeared, human consciousness. Nature is what it is, but without humans, there will be no Humaniverse.

If we ever hope to establish a true human presence in the universe, we must embrace the reality that humankind, not nature, is the cosmos' major player. Human consciousness is wholly new to the universe and must be introduced to it. The Gaia principle subsumes humanity to the needs of the universe. Nevertheless, it will be the universe that increasingly must react to our needs.

Be certain that the human species will do what it has always done—embrace the challenge, vigorously inhale the smell of danger, and seek out ever more intoxicating adventures. We are genetically programmed to incur risks, to seek out the novel, and to accomplish the impossible. However, we must bear in mind that humankind is more than a problem solver. The human is also a creator, an artist—no two terraformed planets will look the same. Our movement around the galaxies will be motivated as much by curiosity as by necessity, our reengineering of planets and solar systems will represent the painter and artisan as much as the technologist.

Over the centuries, these acts of creation will change us forever, expanding our minds, our sensibilities, our imaginations, and our sense of efficacy beyond anything we can currently conceptualize. We will finally see ourselves as the true masters of the universe. In fact, over time we will come to realize that the process by which we create the Humaniverse is not just a way to develop the universe but is also the ultimate mechanism for evolving ourselves. Throughout, as we confront and overcome the physical and intellectual challenges, we will become a better species, smarter and more creative.

As we traverse the universe, it will be the individual, not the group or species that will be ultimately responsible for bringing the Humaniverse to fruition. Yes, the species will number in the trillions. However, let us not permit such large numbers to seduce us into imagining as some do that all of these individuals will congeal into a group mind, reaching a critical mass of brain power that can invent new technologies and create new scientific advancements. While people in groups and teams develop ideas as a result of cross-pollination of ideas, the individual contemplating truth, sans the interference of the media, the Internet, electronics in general, is the greatest source of ideas in the universe. Every great thinker and creator—Tesla, Edison, Beethoven, Shakespeare—ultimately has had to shut out external stimuli and communicate exclusively with their own genius, willing to be persuaded by their dreams. Only then could they return to society and show the others their inventions. Human progress ultimately depends on the individual striving to purposively develop him- or herself. Individuals will find themselves engaged in epic adventures as they conquer time and space. And humanity will discover that it once again requires the emergence of selfless heroes who advance the species. New myths and legends will be created for a new generation of heroes.

To create this Humaniverse, each individual member of the species must dedicate him- or herself to a whole new set of priorities. We are forced to decode nature's most secret rules if only to change them. We are obliged to command the heavens, move mountains, master the forces of electromagnetism, and reengineer our genes. If we do not perform such feats, we will not survive, and the universe at best will remain a lifeless, joyless cavern with no hope of illumination.

8

The Battle for the Future

We live at a time when our species is making dynamic breakthroughs that are setting the stage for a magnificent future. One would imagine that the principal challenges we must overcome to continue such progress would be technological or scientific ones. We would therefore reason that to overcome such barriers we would simply become more scientifically adept and enhance our technological expertise.

However, it is not a lack of technological proficiency that could prevent us from achieving our goals. If anything, our abilities in these areas are accelerating at an almost exponential rate: we are well on our way to taming the atom, controlling our genes, reshaping the earth, and mastering the laser, electromagnetism, and a host of other physical forces.

The formidable impediment that could conceivably prevent the species from achieving its destiny, the barrier standing between humankind and its goals, is *ourselves*, or more to the point, cultural and political influences that threaten to upend scientific progress and choke humanity's advancement.

Not everyone embraces the vision presented in this book regarding humankind's natural destiny. In fact, many vehemently and vigorously oppose any vision that posits that man's destiny involves imposing his will and his consciousness on the world. As we enter the twenty-first century, two major camps in this debate are engaged in a battle for the soul of the species. One side, the expansionary, insists that humankind is destined to aggressively expand human potential and reconfigure the universe. The other side wants to restrict human's activity within predefined parameters—they hope humankind will learn to live "in balance with nature." This battle rages throughout many fields, in academia, in the grade schools, in the media, and in the natural and social sciences. The stakes in this war are extremely high—

if the camp I label "regressionary" wins, I believe that our species will cease progressing, will stagnate, and possibly will disappear.

At the core of this philosophic and political war is the debate over the very definition of *humanness*, and the place of humankind in the cosmic hierarchy. By now you know where the expansionary philosophy places humans: humankind is at the epicenter of cosmic activity, a uniquely gifted species endowed with a special destiny. The opponents of this vision are attacking the idea of the centrality of humanity. They are attempting to define humanity downward, negatively comparing the human species to other entities, including lower primates, "smart machines," even supersapient "aliens" that manifest intellectual and physical qualities that dwarf the capabilities of man. Our own dominant cosmology, the Big Bang theory, conspires to undermine humankind's sense of importance.

These ideas wafting through the culture have real impacts in the political and cultural spheres. Obviously, if we come to believe that humanity is "just another species," no better or worse than other beings (or machines) populating the planet or the universe, we will deny ourselves the right to impose our will on other animals, nature, the biosphere, or the universe. A corollary of this world view is the belief that humankind is a deeply flawed species dangerous to the planet and the cosmos. Many use such suppositions as a justification, a guiding principle, for *political* and *social* activities aimed at thwarting technological innovation and scientific research. Later in the chapter we will examine how powerful forces will use legislative, regulatory, and other means to prevent the introduction of such new technologies as genetic engineering, cloning, the use of nuclear power, and high-speed transport.

The Assault on the Idea of Humankind's Uniqueness

If the human race is to fulfill its destiny to vitalize the planets and beyond, and eventually create the Humaniverse, its members must possess a strong belief in humanity's uniqueness and its special role in the universe. They must envision humanity as a species of unlimited possibilities whose potential is only beginning to be realized. Any ideology that undermines our belief in our abilities, our potential, and ourselves becomes a roadblock on humanity's pathway to destiny.

Unhappily, our educational and cultural institutions now provide the public a view of humanity that is less than complimentary. These institutions promote what I label *equivalency*, the idea that the human species is merely the intellectual equal of a host of other entities. This notion of *equivalency* at one time equates us to other living species, at other times claims that our smart machines such as robots or computers are our superiors. Last, some claim we are the inferiors to that which we cannot see, feel, or even prove. The belief in aliens, creatures from outer space, even the legitimated search for them, has undermined much of our sense that we are the universe's intelligent agent. We can only conclude from such a viewpoint that the human species possesses no unique abilities or rights relative to these other beings.

Animals Are People Too?

Increasingly, a movement is afoot to establish an *equivalency* between the human species and certain other species in the animal kingdom. An organized effort even exists to have certain higher primates recategorized as sapient beings, according them legal rights that traditionally have been delegated exclusively to humans. This movement is gradually and insidiously eroding the distinction between humans and animals.

Humanity has always felt a connection to the animal kingdom—we keep animals as pets, domesticate them, and more often than not become their protectors and stewards. Moreover, humans have always maintained a general sense of fair play in regard to animals—we even have laws against the cruel treatment of the lower species.

The dividing line between fair treatment and exploitation of animals has been a subject of some debate. Some go so far as to stipulate that it is unethical, even immoral, to eat the flesh of animals or wear furs or other articles of clothing made from animal parts. Over the last several decades, the organized animal rights movement, which includes organizations such as the People for the Ethical Treatment of Animals (PETA) and more activist "animal liberation" movements, have become more vocal and strident in their actions on behalf of the protection of animals. Such groups lobby for the rights and protection of animals, some even stage stealth raids on laboratories, zoos, and farms to free select animals from their confinement.[1]

While many seek to protect animals simply because they feel it is the humane thing to do, others have gone a step further. Some people now declare that many species exhibit so many physical and behavior similarities to humankind that such animals must be considered the equivalent of humanity.

As we enter the twenty-first century, this idea of the equivalency of some animal species to humankind is rapidly moving out of the purely philosophical realm into the world of activism. In 1999, New Zealand activists began lobbying that country's parliament to adopt a bill that would confer "human rights" on great apes—chimpanzees, orangutans, and other higher primates. Killing a gorilla would be considered an act of murder, putting a chimp in a zoo a form of unlawful imprisonment. In effect, this bill would redefine the great apes as persons. This hotly debated bill was an outgrowth of the Great Apes Project, founded by scientists, ethicists, and lawyers in New Zealand and elsewhere. Many of the project's scientists have worked with great apes in a variety of field experiments and believe that these apes are too similar to humans to be denied equal rights and status under the law. "We demand the extension of the community of equals to include all the great apes: human beings, chimpanzees, gorillas and orangutans," reads the manifesto of the Great Ape Project. "The lives of members of the community of equals are to be preserved. Members of the community of equals may not be killed except in very strictly defined circumstances, for example, self-defense."[2]

The champions of the bill promote the idea of an equivalency between apes and humans. Their long-term goals include enshrining the rights of great apes in a United Nations Declaration, so that these creatures would now officially be included in the family of man! Biologist David Penny of New Zealand's Massey University, puts it succinctly: "There's now a mountain of evidence that the great apes are as intelligent as young human children." The rhetorical device the signers of the "Declaration on Great Apes" are using is substituting the term *hominid* for humans. In other words, they are not directly arguing that great apes and chimps be specifically classified as humans; rather, they are attempting to get humans classified as part of the larger hominid group. They are willing to even admit that hominids as a class are superior to all other physical species—they claim "the cognitive skills of all hominid species represent a quantum leap from the cognitive abilities of other primates."[3] They just won't accord special status to the human species within the hominid group.

Proponents of the bill, such as primatologist Jane Goodall, cite recent scientific studies showing that humans and the great apes share about 98 percent of the same genetic structure to make the case that all hominids are essentially equivalent. Jared Diamond, professor of physiology at UCLA, has argued that humans and chimps are so similar genetically that it would be more accurate to describe *Homo sapiens* as the "third chimpanzee." They make such claims in spite of the fact that the vast majority of biologists reject the notion that such genetic similarities scientifically prove the equivalence of man and ape. In a recent issue of *New Scientist*, biologists were quite clear on this point: "Genomes are not recipes. A creature that shares 98.4 percent of DNA with humans is not 98.4 percent human any more than a fish that shares, say, 40 percent is 40 percent human."[4]

For the last two or three decades Jane Goodall has been a central figure in this dubious effort to publicize through text and film human-like qualities of the great apes and chimpanzees. Since great apes bully and cajole other apes, have fun, and form political liaisons and other such bonds with each other, they are "just like humans." Others had made similar claims. However, Goodall went one step further, claiming that apes could use sign language to communicate with humans. Even though more recent observational studies reveal that her contentions about chimp use of language were overstated, if not entirely incorrect, the public still believes that apes can talk.

In contrast to the information dutifully reported in the mass media, the case for human uniqueness is overwhelming. In a brilliant book on the subject, *Are We Unique*, physicist James Trefil clearly demarcates the key areas in which humans excel and demolishes the so-called mountain of evidence of the similarities of ape and man. That is, he presents the evidence for the superiority of man.

First, he looks at the contention that Goodall and others have made that great apes and other species can make tools, a skill that we have traditionally considered uniquely human. Chimpanzees in the wild have been observed to take a long stick, strip off its small branches, insert it into a termite nest, and then eat termites that cling to it when the branch is pulled out. Some chimpanzees use rocks to break open nuts. Trefil definitively demonstrates that chimps' primitive use of available implements does not weaken the argument for the singular nature of human tool making. Even though we may think human and animal tool making differ only by "a matter of degree," at a certain point such a gap signifies a major difference between the tool makers themselves. Does anyone imagine that a great ape, a chimp, for instance, could produce a computer, a lunar module, a 747 jumbo jet, the Empire State Building, or develop a science such as genetic engineering? Trefil argues that using a rock to break nuts and constructing a jet engine are so qualitatively different as to be not even in the same category. As Trefil says, although a single raindrop and a raging flood both come from water, they so fundamentally differ they constitute completely separate phenomena. The fact that human technology is qualitatively different from apes' use of sticks and rocks as tools is one reason that the human race, and not orangutans, is reshaping this planet and colonizing the universe.

James Trefil next looks at the contention that chimpanzees use language. According to Trefil, only humans make real language. In spite of what we are led to believe, the experiments to teach various types of chimpanzees language have been less than successful. Goodall's claims that she has taught chimps and gorillas American Sign Language (ASL) have proven to be illusory. Over the years, observers who have visited Goodall's research labs have come away feeling that what Goodall calls "speech" is actually learned behavior in response to visual and other physical cues. In none of these experiments do the chimps show the ability to form grammar, something even any 4-year-old child can do with little training—the ability to link nouns, verbs, and concepts in any meaningful fashion.[5] Goodall and others are engaging in animal training, pure and simple.

Trefil points out the third and possibly most important difference between humans and chimps: only humans possess true consciousness. True, chimpanzees and orangutans, like humans, possess some degree of self-awareness. For instance, some higher apes can recognize themselves in a mirror, even when experimenters have slightly altered the ape's appearance (often by painting part of the chimps' heads red). Nevertheless, true consciousness far exceeds the idea of mere self-awareness. It also encompasses the ability to "know that you know," setting yourself outside of yourself and then looking at your "self" from the outside. Ronald Nadler, a psychologist and emeritus professor at Emory University's Yerkes Regional Primate Research Center in Atlanta, claims that to detect human-like consciousness in apes one would need to learn whether the creature's language allows "thinking about thinking." According to Nadler, "there is no evidence that chimps and the like" possess such a property.

Human consciousness enables us to think at an extraordinarily high level of abstraction, to understand the similarities between seemingly different objects. The quality of consciousness also enables us to shift between abstract thinking and applications of these abstractions to specifics. One of the best examples of this ability to apply abstractions to the concrete is *mathematics*. Humans not only invented mathematics and the theory of numbers, but also apply the same set of numbers to solve thousands of seeming disparate problems, from building houses to constructing computer programs to predicting solar eclipses.

Human consciousness also lends itself to what I label *universal* or *global* thinking. We can see problems outside of our immediate purview, for example, ruminate about the state of our ecosystem, our community, our nation, our world, and our universe. Michio Kaku locates another unique element of human consciousness. Kaku claims that our ability to set our own goals differentiates humans from the rest of creation. We can decide to migrate, invent calculus, fly to the moon. Such lofty objectives are light-years beyond the normal goals followed by other animals to merely adapt and survive—that is, eat and reproduce.[6]

Trefil and others have pointed out that these qualities are due largely to our enormous brain size, about 3 times the size of an average chimp's, and the 100 billion or so neurons that endow us with incredible ability to memorize, analyze, and compute. Goodall, Diamond, and others conveniently avoid reminding us that humans possess this cranial and neurological superiority over apes.

Given the complexity of human consciousness, it would seem almost ludicrous for anyone to preach equivalency between man and ape, let alone try to have it codified as international law. Doubtlessly, some of the animal rights proponents use the argument of equivalency as a mechanism to ensure the protection of the apes from cruelty, abuse, and possibly extinction through excessive hunting. However, these creatures can be protected in any number of less extreme ways. Thomas Insel, director of Emory University's Yerkes Regional Primate Research Center in Atlanta suggests establishing sanctuaries or game reserves to provide "long-term care in a stable social environment."[7]

Many proponents of equivalency would further weaken the concept of humanness. "No doubt some of us, speaking individually, would want to extend the community of equals to many other species as well," admits the Great Ape Manifesto. "Others may consider that extending the community to all the great apes is as far as we should go at present. We leave the further consideration of that question for another occasion."[8] To support this far greater dilution of the definition of human, backers must delegitimize humankind's claim to intellectual superiority or higher consciousness. Peter Singer, author of the 1975 book *Animal Liberation*, claims that if intellectual ability is the main criterion for having rights, then the rights we deny to chimpanzees we should also logically deny to human children with severe intellectual disabilities. If emotions and an ability to suffer are the criteria, then we should grant equal rights to almost any mammal, and probably some of the other higher vertebrates, too.[9]

Granting the great apes the status and rights of humankind will open up a Pandora's box of claims and exemptions for other species. As Frans de

Waal, a primatologist at Yerkes Regional Primate Research Center, explains it, "if great apes were given rights on the basis of continuity with us, rights should be given to other species on the same basis until even rats had them." In fact, in the United States and elsewhere this trend has already started. Increasingly, courts and legislatures are restricting the rights of humankind and expanding the rights of lower creatures of all types, often at the expense of the freedoms and even personal security of humans.

The obfuscation of the differences between man and beast will have a decidedly negative impact on human progress. Clearly, once the population is truly convinced that the human being is not so much *Homo sapiens* but just another ape, a "smart chimp," as it were, we will live down to the expectations of our new "status." Creativity and genius will simply not emerge from individuals who have been convinced that they are the intellectual equivalent of orangutans. One may argue that the Great Apes Proposal and other such movements only seek to raise apes to the level of humans, not lower humans to apes. My response is that the vast majority of the public has a fairly good idea of what an ape actually is, how it lives, and what its limited capabilities are. They can only surmise that by equating apes to men and women we are defining humanity downward!

Establishing in the legal system and the culture itself the idea that apes are our equals will wreak havoc with human progress and our ability to reach our destiny. Why would we desire to establish a Humaniverse and imbue the cosmos with human consciousness and intelligence, if we doubt the innate uniqueness and the superiority of humans? If we have any hope of winning the battle for the future, the educational system and the media must communicate to the public the scientific case for human uniqueness and superiority.

The Cult of the Smart Machine

The concept of the status of the human being as a unique and intellectually superior entity is being attacked from another angle.

The computer has made a valuable contribution to human progress. Through cybergenesis the computer becomes an indispensable tool in our quest to spread ourselves and our consciousness throughout the universe. As we enter the third millennium, however, a handful of cybernetic experts and others are proclaiming that computers, so-called smart machines, will soon become more intelligent than the human being. In fact, some even suggest that these machines will acquire human characteristics, and by implication emerge as the next step in human evolution.

We will examine this emerging debate over whether machines can actually "think," and also consider how an overinflated view of the smart machine can impact the human psyche and by implication human progress.

Can a Machine Think?

Researchers who predict that robots and computers will achieve human capabilities base their contentions on their belief that soon these machines will not only compute but also "think." For decades science fiction novels and movies have featured smart robots with almost human-like thinking abilities. The movies *2001* and the recent *Bicentennial Man* predict a future of thinking machines.

Can the computer, no matter how complex or massive, ever think in the sense that humans do? Such feats as Deep Blue's victory over Kasparov have cybernetic scientists and technicians murmuring that we are on the verge of creating a thinking machine to challenge the human species' monopoly on real intelligence. However, many in the cybernetic community express grave doubts over whether such machines *actually perform human-like thinking.* Marvin Minsky, MIT professor emeritus who is credited with initiating the Artificial Intelligence (AI) movement over 35 years ago, put such proficiency in perspective. According to Minsky, "Deep Blue might be able to win at chess, but it wouldn't know to come in from the rain."[10]

Minsky's comment cut to the very heart of the thinking machine debate. Deep Blue's circuits, wiring, and program, its entire "being," if we can apply such a term to this contrivance, knows nothing except how to play chess. Concepts like "rain" do not even exist in Deep Blue's memory banks. Nor could it even imagine that the rain's overwhelming moisture could impair its circuits, or fashion a strategy to avoid such a misfortune. In addition, skeptics repeatedly remind us that human intelligence created Deep Blue. Yet, instead of celebrating Deep Blue's victory as a testimony to human intelligence, the AI community congratulates the machine for a job well done.

Actually, this debate has already been settled in favor of humankind. Cambridge University physicist Roger Penrose combines information science, cognitive psychology, and physics to make a tightly constructed case against the possibility that computers can ever achieve human intelligence. In two books, *The Emperor's New Mind* and *Shadows of the Mind,* Penrose argued that the computer can never be conscious, and thus truly intelligent.[11] When our brains operate, we juggle many different thoughts and thought patterns before zeroing in on one unified pattern that becomes a conscious thought. Some physical mechanism must exist that helps us achieve, and maintain, this pattern of multiple simultaneously existing "protothoughts" before we focus in on the final thought. Penrose claims that this mechanism acts "non-locally." That is, some aspects of these thought patterns would have to act more or less at the same instant *at widely separated locations of the brain*, rather than spreading out relatively slowly, neuron by neuron.

The genius of Penrose's theory is the way he applies quantum physics to the operation of the brain. His basic point is that before a thought, or the neural signals that constitute thought, enters our consciousness, it exists in a "quantum wave state." At the threshold of consciousness, the "wave-thoughts" then "collapse" or coagulate into a single ordinary thought.

If, as Penrose claims, such quantum mechanical phenomena are the operating principles behind human consciousness, the brain functions in a way that no mechanical device, computer or otherwise, can ever replicate. Computing devices, artificial neural networks, cannot simulate quantum mechanical phenomena.[12] Penrose's theory seems to prove that no matter how complex or sophisticated a computer or computer network is it will never achieve consciousness. And if our smart machines can never reach consciousness, they will never be said to truly think![13]

Donald Norman, VP of research at Apple and psychology professor at the University of California does not believe that in the foreseeable future computers and robots will come to mimic and/or surpass people. People and computers operate along completely different principles. According to Norman, the power of biological computation, that is, the human brain, emerges from "a large number of slow, complex devices—neurons—working in parallel through intricate electrical-chemical interactions."[14] All this hardwiring enables the human to think in amazingly complex, abstract ways. On the other hand, computers have no problem finding square roots instantaneously, or adding large columns of eight-digit numbers without hesitation. The computer's ability to perform math with ease and dexterity results from its multitudinous, high-speed devices following binary logic. Errors in the operation of any of the underlying components are avoided either by careful design to minimize failure rates or through error-correcting coding in critical areas.

Because of the computer's speed and accuracy, Norman says, we accord computers positive traits such as precise, orderly, undistractible, unemotional, and logical and label humans vague, disorganized, distractible, emotional, and illogical. According to Norman, we have our priorities backward. To appreciate humankind's natural superiority, he says, let us label humans creative, flexible, attentive to change, and resourceful and stigmatize computers as dumb, rigid, insensitive to change, and unimaginative.

Can a Machine Discuss Poetry?

Such arguments have not deterred some futurists from predicting that machines will become as smart as humans; after all, our computers and supercomputers are getting faster, larger, and more powerful. Some theorists even predict that this growth in computer power could mean that computers, not humans, are the next stage of evolution on planet Earth.

In his *Age of the Spiritual Machine*, Ray Kurzweil predicts that by 2099 these machines will become so intelligent that "there is a strong trend toward a merger of human thinking with the world of machine intelligence." Remember that throughout the twenty-first century Kurzweil has us connecting to the neural net just to get along and do our jobs. Now, in 2099 no longer does any clear distinction remain between humans and computers. He says that although humans still exist, "the number of software-based humans vastly exceeds those still using native neuron-cell-based computation." And those still of "carbon-based design" use neural-implant technology, which provides extraordinary augmentation of human perceptual and cognitive abilities—if one does not use such implants, one is out of the loop, "unable to meaningfully participate in dialogues with those who do."[15] What about our bodies? "Most conscious entities do not have a permanent physical presence." Each of us will be able to assume any number of virtual presences whenever we desire. However, the body is no longer a functional aspect of the human being.

Robot expert Hans Moravec, whom we met earlier, also predicts that humans will eventually disappear. In his scenario, though, we replace ourselves with this wisely wired new entity—the universal robot. As mentioned, Moravec has robots attaining human levels of intelligence by 2040. By 2050, they will have far surpassed us. Since the robots are our evolutionary heirs, Moravec thinks we should consider them as "children of our minds." In the last stage of transition from humans to robots, many of our human descendants, our flesh and blood grandchildren, decide to become "exhumans," by uploading "themselves," or some computerized version of their minds and personalities, into advanced computer-robots. By doing so we become immortal, living forever as these robots. Others in the physics and cybernetic community mirror Moravec's prediction that we will develop sentient, thinking robots. Michio Kaku speculates that by 2050 some form of thinking worker-robot will be available. He does not go so far as to predict that we will meld into these machines.[16]

Smart-machine advocates rarely mention the fact that in their future this merging of man and machine really represents the submerging of man into machine. Peter Cochrane, head of British Telecom's research efforts to invent the strange Soul Catcher neural implant, portrays this melding of machine and human as what he considers paradise on Earth. Others may find Cochrane's ethereal future less than inviting: "Ultimately . . . perhaps even the merging of all human life forms into a single ethereal being supported by an organic hardware with both growing and developing like some cosmic jellyfish." He enthuses that once we develop the global neural net we can be everything, be everywhere, be everyone, experience everything! Not alarming if Cochrane was England's reigning poet laureate, but an unsettling statement when it comes from the research director of the United Kingdom's most powerful telecommunications giant.[17]

Although Kurzweil insists that his global neural net is the next stage in the development of the *human* species, his later words betray him. To Kurzweil, humanity and human intelligence are not entities that ought to be evolved, only mechanisms "providing the means for the next stage of evolution, which is technology." He goes on to say that the emergence of technology is "a pivotal event in the history of the planet." So the smart machine crowd, much like the New-Age/Teilhard enthusiasts mentioned earlier, believes that the real entity to be developed is not humanity, but the Earth. In their scenarios, the human species is a mere step in the process of Earth's evolution. Moreover, Kurzweil even warns us not to interfere with the development of cybernetic technology! As Kurzweil explains, "The introduction of technology on Earth is not merely the private affair of one of the Earth's innumerable species." That's the human species to which Kurzweil is referring.[18]

The melding of the human race into this global neural net most certainly dooms our species. The virtual reality ghosts in the supermachine will soon suffer the same fate their physical counterparts did. They will follow us into cybernetic Valhalla, eventually ceasing their individual existences. From the neural net's perspective, like their physical predecessors, they will serve no purpose for a machine that just wants to continue existing in a universe of perfect computation.

Why We Must Embrace the Human Future, Not a New Machine Age

Our innate belief that we are the end product of evolution, or at least the highest form of life and intelligence on the planet, provides us with the confidence and self-esteem that fuels our relentless quest for knowledge, our conquering of disease, our establishment of dominionization of this planet. Conversely, if humanity accepts the idea that the smart machine is not merely an extension of human intelligence but in some way is an advancement over the human species itself, progress in all areas, including cybernetic science, will slow or grind to a halt. The reasons why are not difficult to discern.

For one thing, our belief that the machine is superior to humans will lull us into expecting the machine to perform operations beyond its abilities. We will expect the smart machine to not only do our computation and calculations, as it does now, but also perform operations for which it is not prepared, such as establishing a purpose for a project, solving problems, and generally operating at a high level of abstraction. The computer can mimic some human faculties, but don't ask it to run your government.

Although the computer plays chess, runs transportation systems, and guides our satellites, it always does so ultimately under human guidance and supervision. In addition, humans have always had to intercede to push computers to their next level of development, and always will. Human ingenuity will create the next generation computers to run our increasingly complex transportation, energy, and communications systems. We court disaster if we shift this responsibility for creating higher machine intelligence to the machines.

Another tenet of the smart machine ideology, if accepted, will have profoundly deleterious effects on human progress. Smart machine cultists believe that smart machines, not humans, ought to be the primary explorers of the cosmos. Currently we envision robots serving as reconnaissance agents for human expeditionary forces, examining the terrain and sending back pictures and data useful to future human crews. The smart machine crowd has other ideas—in their future robots or other machine formats will become the true, no, the only, astral pioneers. Hans Moravec imagines that smart robots, human-like machines, will not only explore the universe but will eventually populate it.

Ray Kurzweil explicitly states that intelligence, however defined, should play a role in the development and perfection of the universe. However, he does not predict that this intelligence will be *human*. He has trapped the species into this ethereal neural network, and does not make clear how this intelligence coursing through the neural net will ever make it off this planet, let alone traverse the cosmos. British Telecom's Peter Cochrane thinks he has an answer. We will enjoy "instantaneous teleportation by radio and light beams. Beyond the moon in 2 seconds, the sun in 8 minutes, the galaxy in 100,000 years." Cochrane says we might become as "a wave of energy propagating throughout the universe looking for a host machine or entity." Alternatively, perhaps we would meld with other intelligent machines or beings of a "similar nature." Cochrane thinks this could lead to "the very creation of the ultimate life force—a oneness throughout."[19]

These scenarios are hardly a desirable human future from the expansionary perspective. If we ever shift sole responsibility to these smart machines for the exploration of the universe, vitalization as defined earlier will never occur. The reason for this lies in the nature of the exploration and vitalization processes themselves.

First, let's deal with the obvious objection to the machine taking over space exploration. The very definition of vitalization amounts to no less than the *human* species extending its humanness, which is a combination of various cognitive and affective aspects of our being. We will build rational systems and make the universe behave more predictably. In addition, we will export our serendipity, the wild creativity that infuses our music, art, and architecture. The yearnings of smart-machine enthusiasts' arguments aside, machines simply do not create beauty without human input. Let us not permit the computer's ability to generate virtual images, mimic songwriting, and otherwise seem "human" seduce us into believing that the smart machine will ever develop the true consciousness, the real creativity, that the universe craves. Machines will help us chart the cosmos. But they can never vitalize, because *they are not alive.*

Another reason why humans, not machines, must become the explorers of the universe can be found in a distinctive complementarity between a property of the universe and human nature itself. As we saw in our description of the vitalization process and in our discussion of the expansionary view, outer space presents the explorer with a plethora of unexpected challenges. Space is an agglomeration of ambiguous phenomena—black holes, dark matter, the tricks of space-time, multidimensions—that will surprise, fascinate, amaze, befuddle, and at times frustrate the being who encounters them.

The human species may be the only entity in the universe, and is certainly the only entity on Earth, that can deal with the ubiquitous serendipity that is outer space. The explanation lies in our basic biological and cerebral makeup. According to Donald Norman, the human's biological computation process itself is inherently error tolerant; our actions—unlike the output of computers—are fraught with imprecision and inaccuracy. Ironically, when dealing with puzzling systems such as space/time, these tendencies become a strength, not a weakness. Norman states that "humans have developed systems that can handle ambiguity. In fact, in many cases they were forced to do so in order to survive as a species." Moreover, this ability to deal with ambiguity has helped us achieve dominionization over matter and the Earth. And on our space voyages this quality will help us counter nature's surprises, adapt to them, and eventually reshape them into something more beautiful, more ordered, more perfect.

Humans, not robots, will dominate the future, invent, create, pursue dominionization, and continue to improve the planet and enhance our civilization. True, the cybergenesis process will enhance the human being through the addition of new and improved body parts and implants. However, even with all this tinkering the human species will still retain its intrinsic essence.

The real challenge will not come so much from smart machines as from those who try to convince us that the machine is outstripping humanity. Such misinformation, if allowed to fester with no rebuttal, will erode the self-confidence we as individuals need in order to muster the energy and will to spread human consciousness throughout the galaxies. At the 1999 World Future Society (WFS) Conference in Washington, DC, only a few months after the publication of Kurzweil and Moravec's books, several presenters were already heralding as "serious" and "groundbreaking" concepts like the "global brain" and the "robotic future." *Futurist* magazine ran as a cover story Kurzweil's article on the "spiritual machine." The WFS membership represents the cutting-edge cognoscenti of a host of fields, including science, academia, politics, government, even religion. Ideas that gain acceptance among the WFS crowd soon fan out into the general culture.[20]

Over the coming decades, the future of the relationship between human and machine often will be debated. In such debates the human species must be defended by those who recognize the intrinsic worth of the human being and accept the responsibility engendered in our very existence. For the other side to triumph it will have to explain the seemingly contradictory idea that the species that designed the smart machine can no longer compete with its own creation. I believe that at the end of these great debates we will come to believe even more in the intrinsic uniqueness of the human species.

Alien-ation

A third idea undermining the public's faith in the concept of human uniqueness is the growing belief in the existence of beings from other worlds. This
phenomenon includes a variety of experiences and claims: UFO sightings,
"close encounters" with aliens, even forced abductions of humans by extraterrestrials. The proponents of the existence of extraterrestrials present us
with the image of beings vastly superior to us technologically, physically, and
intellectually. Sometimes malevolent, other times inquisitive, these extraterrestrial visitors invariably exhibit the capacity to exert power over humans;
rarely can we control them.

Belief in the existence of UFOs has grown in recent years. A recent Gallup
Poll found that the percentage of college graduates who believed that flying
saucers or other such crafts from distant worlds have visited Earth jumped
from only 30 percent in the 1970s to 42 percent in the late 1990s. The belief
in UFOs, aliens, and other space mythologies has gone mainstream—a 1997
Gallup poll found that 71 percent of Americans think the government knows
more about UFOs than it lets on. A Scripps Howard News Service poll
released at the beginning of 2000 found that almost 60 percent of Americans
believed that some time in the future humans will discover space aliens.
Predictably, Americans surveyed predicted that such aliens will be "superior
to humans."[21]

The media does little to dispel such beliefs. The wildly popular TV program, *The X-Files*, movies such as *Men in Black*, the cottage industry of books
purportedly proving the existence of UFOs, and the publicity accorded supposed alien abductees, all give credence to such a belief. On July 4, 1997,
more than 100,000 UFO enthusiasts gathered for a weekend-long nationally
televised event dedicated to the subject of UFOs, alien visitations, and the
like.[22] In December 1997, the First World UFO Forum was held in Brasilia.
Thousands attended this conference to hear 50 internationally renowned
researchers from dozens of countries present papers on phenomena such as
alien abductions, sightings, and other forms of close encounters (including
stories of humans who claimed to have had sex with aliens).[23]

Although traditional science scoffs at the idea of alien contacts and UFO
sightings, scientists' statements regarding astronomical discoveries ironically
often encourage such beliefs. For instance, recent discoveries of distant solar
systems, replete with planets, led some scientists to publicly speculate that
life in some form must exist somewhere in the universe. Science tells us that
in this solar system alone, there are at least four spheres besides Earth where

life could have evolved. At a recent meeting of the American Astronomical Society, Stanford University scientist Chris Chyba was so taken by the discovery of new planets in our galaxy that he claimed that "There is a substantial higher optimism for the existence of life beyond the Earth." Chris McKay, a planetary scientist at NASA's Ames Research Center, was quoted at the conference as saying "Even if there is only one star in a thousand that could harbor life, that's still a billion stars in this galaxy alone."[24]

The fact that respectable scientists are actively searching for signs of life in the cosmos also encourages popular belief in the existence of aliens. The late Carl Sagan became an ardent supporter of the search for extraterrestrial life forms.[25] Today there are numerous SETI (Search for Extraterrestrial Intelligence) programs in operation, hunting for some manifestation of any alien civilization's technology by analyzing weak radio signals from space. In addition, the University of California at Berkeley's Astronomy Department has just established the Watson and Marilyn Alberts Chair for the Search for Extraterrestrial Intelligence—the first academic chair of its kind anywhere. The first professor to hold the chair, William Welsh, will collaborate with the SETI Institute to build an array of five hundred to a thousand commercially available satellite dishes, like your backyard dish, and connect them so that they become a very large and sensitive radio ear.[26]

What is the chance aliens exist and that they have visited the third planet from the sun? About the same as a great ape composing a choral arrangement for Beethoven's Seventh Symphony. At the 100th annual meeting of the American Physical Society, Lawrence Krauss, chairman of the physics department at Case Western Reserve University in Cleveland, stated that "Extraterrestrial aliens haven't visited here and won't be coming here." Like most physicists, Krauss's skepticism derives from the law of probabilities and laws of physics. Even if advanced civilizations exist elsewhere in the universe, the same laws of physics bind them. These other species, if they do exist, are most likely trillions of miles from Earth. The limits of time and energy preclude the possibility of them traveling directly from another galaxy to this one.[27] In addition, the chance that this intelligence evolved into something advanced enough to fly over the light-years of space to find Earth is even smaller. Many scientists flatly reject the possibility intelligent life exists *anywhere* in the universe other than Earth. UCLA astronomy professor Ben Zuckerman, a veteran of numerous debates with SETI scientists, states that "I've pretty much reached the conclusion that the occurrence of technological life is an extremely rare occurrence."[28]

This is not to suggest that UFO sightings are hallucinations. Actually, most people who sight a UFO are simply observing an entity they cannot name. For decades, many who reported UFO sightings were accidental witnesses of top-secret government experimental combat and surveillance aircraft. For example, as far back as the 1930s the U.S. government, as well as the German government, were designing and testing prototypes of saucer-shaped vehicles. Such vehicles could fly at extremely high speeds but stop quickly, perfect qualities for landing in confined areas, such as aircraft carriers. By the early 1940s, the U.S. Navy had built a prototype of a saucer-like "Flying Flapjack," the secret XF5U-1 aircraft. By 1947, the year of the first UFO sighting, the Navy could have developed a fully operational vehicle of this type, especially since the U.S. government could have incorporated into the craft's design recently "appropriated" saucer-based flight technology from the Germans.[29]

What then can we make of the thousands of stories people tell of being abducted by aliens? According to psychologist Norman Remley of Texas Christian University, "most of the people who report such experiences have a previous awareness of aliens. They have seen television shows, movies, read books and talked to other people who say they've been abducted." Since most of these abductions occur while the subject is asleep, scientists have concluded that the conditions under which these reported abductions occur resemble a form of *sleep paralysis*. The subconscious of the abductees has incorporated a media-driven image of aliens, which they then generate in their dreams. They awake from their dream, and sense that someone, something, is present in their room—they may even hear buzzing sounds and see bright lights. Although genuinely frightened by these phenomena, they cannot move or run to escape. This experience is so real for many of the subjects they will continue to believe it occurred even after third parties inform them that they actually saw the subject physically sleeping in his or her bed at the time the abduction supposedly ensued.[30]

Even if the "abductees" produce no physical evidence, their harrowing stories of alien kidnapping convince many listeners of the tales' veracity. One psychologist, Harvard professor John Mack, after investigating the experiences of hundreds of abductees eventually came to believe his subjects' claims. In his book *Abduction*, Harvard's Mack declares that "we may be witnessing an awkward joining of two species, engineered by an intelligence we are unable to fathom, for a purpose that serves both of our goals with difficulties for each."[31] Unfortunately, the constant stream of stories about UFOs, alien sightings, and abductions that appear in books, movies, documentaries, docudramas, and on Web sites has convinced a large proportion of the population, much like Professor Mack, that aliens exist.

Of course, our main concern here is the impact such beliefs in aliens have on our image of ourselves as a species. Clearly, the belief in supersapient visitors from the cosmos is potentially as deleterious to our self-esteem and self-image as our acceptance of the contention that we are equal or inferior to apes and smart machines. In other words, the myth of the supersapient alien is another theatre of combat in the battle for the meaning of humanity.

A particularly insidious aspect of the belief in supersapient aliens is that their defenders, their proponents, can imbue these creatures with whatever superhuman or supernatural qualities they so choose. Since none of us have ever experienced, nor will ever meet, one of these Martians or Venusians, the abductees are at liberty to portray these creatures as possessing physical attributes and technological capabilities far advanced to our own without fear of contradiction. In short, the 50 to 70 percent of the American population who say they believe in the existence of aliens must also accept as true our visitors' technological and cultural superiority over *Homo sapiens.* Put another way, a sizable group of Americans believe that in spite of all these eons of hard work, creativity, sacrifice, and suffering, the human species is at the bottom of the cosmic hierarchy.

Such beliefs undermine our regard for the human being. For one thing, by subscribing to the idea that "they" are among us we can no longer believe in the uniqueness of the human species—others in the galaxy possess higher consciousness. Once we reject the idea of our uniqueness, we can no longer maintain our belief that humanity is endowed with a special destiny. After all, our alien visitors are already doing what we assumed our unique destiny would require us to do—traipsing around the heavens, performing feats of technological wonder, reengineering the universe, establishing beachheads throughout the galaxy. Even if we ever achieve their level of technological efficiency sometime in the future, at best we will be repeating feats they have already accomplished. Furthermore, we are no longer required to perform these magnificent acts. The aliens, after all, are doing it for us.

One little-noted aspect of the alien phenomena also undermines our self-image. Our friends from afar could if they so wished help us upgrade our current technological skills to reach their level of sophistication in the sciences—we could certainly use some pointers in astronomy, aeronautics, and medicine. Yet over the last half-century these cosmic slummers have acted positively anal when it comes to sharing information we could use in such areas as biogenesis or dominionization. Their tendency to hoard scientific secrets conveys a terrible message about their attitude toward us. They are clearly not requesting, nor do they require, our assistance in perfecting the cosmos. Alas, we can only conclude that our intergalactic visitors consider us unfit as potential allies; from their perspective, we are at best specimens. How inferior to these extraterrestrials are we, anyway?

According to our liaisons to the alien world, the abductees, the last thing the otherworldlies want to provide mankind is scientific knowledge. After all, as soon as we learned to split the atom we used this knowledge to build and use atomic weapons. These aliens appeared in 1947, just in time to monitor our species to ensure that we did not use our growing technological capability to do serious damage throughout the universe. The hidden message is that humankind is an innately pernicious species that must be protected from its worst impulses.

This alien myth is more noxious than the arrogant claims of either the Great Apes Project or artificial intelligence hucksters. We can empirically prove or reject ethnologists' claims that apes can learn and use language as humans do. When computer scientists pronounce that machines can think, we can compare humans and machines to judge the viability of such claims. By comparison, we cannot scientifically repudiate the claims of extraterrestrials' preeminence, because all the "evidence" related to aliens is insubstantial—anecdotes, hearsay, and "testimony." No videotapes, no verifiable transcripts, not even a good parlor trick. No physical evidence, no examples of superior technology, and worse, no ET!

What "they" have left us, however, is something much more powerful than a working model of their spaceship. A public that accepts the existence of aliens that can travel faster than us, capture us, and probe and manipulate us at will is a public slowly acquiescing to the idea of humanity as a second-rate species. Although these phantoms can cause us no physical harm, cannot invade our borders, or capture our cities, our persistent belief in their existence will eventually subvert the material and cultural progress it took centuries for us to achieve. Even fantasies can be real in their effects!

At root, the ET phenomenon signifies that the human species is apprehensive about the potential and the responsibility of their newfound powers. We, not Martians, are becoming "superhuman," controlling forces of nature that until recently seemed to be in the domain of the gods. However, we react to our newfound power like adolescents who all their lives hunger for their "freedom," and then after graduating college mysteriously invent excuses to continue living with their parents. How much more comfortable humankind finds it to be coddled by our friends from afar then to accept the moral responsibility to undertake the arduous task of vitalizing the universe and reversing the forces of entropy.

In a recent interview, I was asked why people persist in believing in the existence of extraterrestrials. I replied that people feel that "these aliens will be able to upgrade our lives." I labeled this cultural phenomenon a form of "wish fulfillment."[32] However, I also remarked that over time we would outgrow this need to believe in the existence of extraterrestrials. This will happen as a greater proportion of the population contributes to the scientific enterprise and learns to operate and control these new technologies. As each of us invents, creates, and helps the society and the universe move forward, we will come to understand that there are supersapient beings that walk among us. They are called humans.

Only when we shed the image of ourselves as inferior beings in the universe's "class system" will we be prepared to undertake the responsibilities we have ahead of us.

Cosmology Counts

Many imagine that the picture that each of us maintains in our minds about the way the universe works, how it began, and how it will ultimately end, while perhaps interesting and intriguing, is an abstract principle having little practical impact on the way we live our lives.

In reality, our image of the cosmos significantly influences our actions, thoughts, intentions, and life goals. Moreover, this image will determine whether or not the species ever achieves its great destiny. In other words, cosmology counts!

While observers have complained that the average person learns only a smattering of science, every school child has absorbed a core set of basic facts and major images of the way the physical world operates. One of these kernels of scientific knowledge is the Big Bang theory. Most high school graduates can recite the major tenets of science's creation tale—the world, the universe, was initiated in a big explosion, almost from nothing. From that explosion emerged all matter, which then dispersed outward at fantastic speeds and eventually formed the galaxies, stars, and our home planet.

The end game our schools teach most resembles the Big Crunch scenario—at some undefined time in the future all matter will be dragged back by the forces of gravity into a compressed heap. (Most schools and universities are not yet teaching "the eternally expanding universe" version of this story, the one that ends in the Big Chill.)

Although alternate theories to the Big Bang exist, the Big Bang theory is presented to students and the general public as the one true cosmology. The Big Bang theory's dominance is such that it has become the central frame of reference for most academic and scientific disciplines. In effect, as physicist/cosmologist Eric Lerner says, the underlying current of our lives has become a vision of the universe on a one-way street from an "an explosive start to an ignominious end." We all believe in a universe in decline, after 20 billion years of existence now actually running out of steam. We are told that any progress or evolution we detect in nature and human society is at best an accident, providing an illusion of advancement within a universe whose modus operandi is decline and annihilation.

What is the impact that such a pessimistic cosmology might have on society and culture, and eventually the quality of the lives of individuals in these societies? In his recent book, Eric Lerner describes the enormous influence the Big Bang view of the cosmos has on science, culture, the arts, and ultimately on economic and material progress. Lerner directly traces back to the Big Bang theory the pessimism that permeates the arts, the media, and pop music, the seeming lack of progress in many scientific fields, and the inability of many economies around the globe to reach significant levels of growth. Social critic Gregg Easterbrook largely concurs. "In part because it has been assumed that science would inexorably prove existence to be no more than a chance manifestation of pitiless mechanical forces," Easterbrook opines, "the main current of postmodern thought in philosophy, literature, art, and their mass-cult equivalents has been silted with gray."[33]

The influence on current social trends of cosmologies predicting the triumph of entropy becomes clearer if we examine the nature of this postmodern philosophy. Social commentator Frederick Turner defines *postmodernism* as a theory that rejects the modernist ideals of progress, virility, rationality, coherence, and the unity of artistic form. It is quite logical that a culture that has absorbed the Big Bang theory into the deepest recesses of its consciousness would adopt postmodernism as its dominant ethos. If the universe as we know it is inexorably heading toward oblivion, why should our culture embrace progress, even as an idea? Virility implies taking aggressive actions intended to impact and transform nature and the universe. But why bother perfecting a universe that the Big Bang theory assures us will eventually acquiesce to the forces of entropy? Moreover, why endorse coherence as a working principle for society and try to embody this concept in our artistic creations, if the universe will eventually tear itself to pieces or recompress back to its original state, a single atom?

Turner says that postmodernism also rejects artistic genius, psychic integrity, individualism, and the ideal of the enlightened state. It is clear how the current cosmology would encourage the rejection of such values. The artistic genius looks to make sense of the world and elevate it to a higher state. However, he creates in vain, the modern cosmologist declares, since the cosmos will eventually grind his highest achievements into dust. The enlightened state tries to elicit social order from chaos, and redirect peoples' energy to achievement of lofty goals. However, in the Big Bang universe, the government that tries to create progress is only forestalling the inevitable, the ignominious end promised by the Big Bangers.

We would also expect that postmodernism, rising from a culture whose cosmology predicts total destruction, should champion and reward a commitment to disorder in the arts. The less sense an artist makes, the more in synch he or she is with the ultimate reality of the Big Crunch.[34]

In one of his last writings, Carl Sagan, viewed by so many as the "eternal optimist," betrays the postmodernistic tendency to denigrate the human species, its efforts, and its output. Sagan tries to determine if there is anything about human activity at the turn of the millennium that would qualify current society as unique. He admonished us not to vainly think that we are reaching a "critical stage of development." "There is only one sense in which we are special," Dr. Sagan expounds. "Due to our own actions or inaction, and the misuse of our technology, we live at an extraordinary moment for the Earth at least—the first time that the species has become able to wipe itself out."[35] That's it! All these centuries of work and sacrifice have finally brought us to the point where we can destroy with abandon all we survey.

As the despair-ridden Big Bang cosmology seeps into the culture, pessimism becomes a backdrop to much of human activity. Pop culture in the West during the 1990s, especially in its music and movies, glorifies violence, Satanism, sado-masochism, death, destruction, and degradation of women. As we enter the third millennium we observe an increasing proportion of the population exhibiting a marked inability or an unwillingness to frame a clear vision of the future. Many American and European youth consider it a badge of honor to endorse irrationality as a posture and a value. In Germany, the United States, Italy, Japan, and elsewhere, fertility rates have dropped to the point where many of these nations' populations will soon begin to shrink. Recent censuses have revealed that a large percentage of young adults in the United States no longer choose marriage as an option. One-third of generation Xers, those between the ages of 25 and 35, and 21 percent of the population as a whole, have never been married, a trend the U.S. Census predicts will only worsen by 2010.[36]

Another aspect of the Big Bang theory that has an unsettling effect on the culture is its proclamation that creation itself was a purely random occurrence. In such a world the appearance of the planets, the galaxies, and even life itself are just products of happenstance. Writer Adam Wolfson remarks that the Big Bang cosmology's inherent randomness represents a major break from our centuries-old belief, rooted in Judeo-Christian traditions as well as Enlightenment philosophy, in an ordered cosmos governed by providence or a fundamental rationality. Wolfson thinks that the worldwide spread of gambling, especially pure games of chance like the lottery, is directly linked to our embracing of cosmologies that explain the world as a kind of chaos. Gambling has become an obsession, a major industry, in America, the West, and most of the developed world. By 1996, 10 states in the United States had casinos, 36 had lotteries, and 24 allowed Indian gambling. Dozens of countries plan to open major casino resorts to try to corral some of the international money going to Las Vegas and Atlantic City in the United States.

In a world that accepts the Big Bang cosmology, creation itself "is perceived as a cosmic accident, and man's existence as a fluke of evolution."[37] People reason that since the creation of the universe and life, as well as the evolution of life into humankind, is considered just a fluke, why not play Powerball or State-run lotteries. As Wolfson asks, "Were the chances of mankind and the world coming into existence any greater than 80 million to 1? Probably not. In which case the odds of winning Powerball look pretty good."[38]

While this "cast your fate to the winds" mentality might be fine in primitive cultures, he says, it can prove devastating in a society that requires planning and rational decision. Advanced industrial society will find itself in trouble if it ever becomes infatuated with the concept of chance and luck as the ruling *Weltanschauung* of the age.

One of the supreme ironies is that for all its profound impact on our society and culture, the Big Bang may not even be scientifically valid, a possibility we considered at the end of Chapter 6. In his book *The Big Bang Never Happened*, physicist Eric Lerner asserts that data from recent astronomical observations from the Hubbell telescope and other instruments cast doubt on the theory that the universe originated in a single cataclysmic explosion. Lerner thinks that "plasma cosmology," developed by Swedish Nobel laureate Hannes Alfven and others, offers a better explanation of our physical universe. The plasma theory posits a universe awash in plasma, a gaslike substance consisting mostly of electrons torn from atoms that is influenced more by magnetic fields than gravity. On Earth, the most dramatic forms of plasma are lightning and aurorae. The sun and the stars are gravitationally bound spheres of plasma; on larger scales, plasma gas has been detected at the center of the Milky Way and pooling around radio galaxies.[39] Unlike the Big Bang theory, the plasma model holds that the universe will go on forever, changing shape and increasing in complexity. Lerner admits that much empirical research is needed to confirm the plasma model's validity and flesh out its parameters.[40]

Others besides Lerner question the validity of the Big Bang theory. One group of physicists declared in a 1990 *Nature* magazine article that the Big Bang theory was so deeply flawed that it did not deserve its current status as "the correct" model of the universe. These physicists, representing such notable institutions as the University of California and the Max Planck Institute for Astrophysics, proposed an alternative "steady-state" model of the universe. In this model, the universe has no spatial or temporal beginning or end—matter is continually created via a succession of "little bangs." New galaxies form constantly, at a rate largely conditioned by the speed at which the universe is expanding.[41]

Certainly, a revision of the Big Bang theory may be in order. The new theory that might replace the Big Bang does not, however, have to reject the idea of an *evolving* or changing universe. Such concepts seem possible within the plasma or steady-state models. Moreover, the theory's demise does not signify that what we understand about the establishment of the conditions for the emergence of planets, life, and humanity has to change. It just means that we will have to divest ourselves of the idea that the universe emerged "out of nothing" and also shed the idea that the universe will end in a Big Crunch.

However, let us assume for the moment that the Big Bang cosmology is the valid explanation of the way the universe came into being, as the majority of scientists contend. The expansionary vision of human development posits that through human effort our species will create the Humaniverse, a structure, an entity, that is imbued with human consciousness, order, and rationality. As we evolve the Humaniverse, the human species will be enhancing its mastery of the laws of physics and nature. Eventually we will be applying this new knowledge to improve and ultimately perfect this entity. Even if this theory is correct, human intervention can prevent the horrors predicted by the Big Bang theory, including the heat death. Human life challenges even the so-called Second Law of Thermodynamics. Therefore, it seems totally logical that, millions of years into the future, we will develop the power to overwhelm the ultimate expression of this Law, the Big Crunch or the Big Chill, and reverse the fortunes of the universe. In fact, if the Big Bang theory is actually correct, our presence becomes even more crucial and our mandate stronger.

New world views are rarely welcomed onto the scientific stage. We can just imagine how the scientific *cognoscenti* would greet an expansionary view of human development contending that humanity through sheer willpower can shape the ultimate behavior of the universe and through "purposive self-development"and fashion its own advancement.

Nevertheless, this world view must prevail! Only when society adopts a vision of the human species as a creative force in the cosmos will it be able to rid itself of the pessimism permeating the culture. We must also recognize the existence of alternate cosmological models, such as the plasma or steady-state theories, which present us a universe evolving from an infinite past to an infinite future. Moreover, our schools and universities must teach that even within the Big Bang theory lies a more reassuring "happy ending," Dyson's benign "precisely critical" universe, in which no Big Chill or Big Crunch awaits our descendants.

The public must be made aware of new findings in science suggesting that the universe encouraged humanity to emerge. Let our institutions also inform the public that whatever the true picture of the Earth's origins and evolutionary trajectory, it will be humankind that determines its eventual development.

The Battle over Growth and Scientific Progress

By their very nature, the forces dominionization, biogenesis, cybergenesis, and species coalescence require that the human species fundamentally transform nature. We create new energy sources, manipulate the very basic building blocks of matter, and defy gravity as we prepare to travel to and colonize distant spheres.

As we have seen, a host of cultural influences, especially those that define humanity downward in the hierarchy of being, conspire in many ways to infuse humankind with self-doubt about its right to pursue such activities.

Let us not minimize the impact that such ideas would have on our future should they predominate. Once humankind perceives itself merely equal to or slightly inferior to machines, animals, and our friends from other galaxies, we will also become predisposed to believe that we have limited rights vis-à-vis entities such as trees, forests, and the "ecosystem." Therefore, we will begin to perceive that activities relevant to vitalization might be violating the rights of nature.

These cultural ideas have spawned and simultaneously provide justification for a host of social and political actions and movements that look to restrict human expansionary activities. A perfect example of a mechanism to subvert human progress is the move to pass highly prohibitive national and international laws in response to "global climate change" and global warming.

The global warming hypothesis claims that "irrefutable evidence" indicates that human activity, mainly industrial output and automotive exhaust, is causing Earth to get progressively warmer each year and that this temperature rise will have devastating consequences for the planet. The carbon dioxide, methane, and other gaseous emissions from factories, cars, and other sources will collect in the upper atmosphere at such a rapid rate that they will trap in our atmosphere much of the heat that would ordinarily escape into outer space. Over the next century, this trapped heat will cause the planet's temperature to dramatically increase. As a result, glaciers will melt, sea levels will rise, coastal regions worldwide will become submerged.

Much to the average reader's surprise, legions of reputable scientists see little evidence that global warming is occurring. True, over the last century, temperatures worldwide have gone up slightly. However, most of this increase ended about 40 years ago, when the level of industrial activity, and hence emission of gases, was much lower than it is today. In fact, over the last several years, global temperatures have actually decreased. As we perfect our instruments that measure climate patterns, we find it more difficult to predict any future global warming whatsoever. The UN's Intergovernmental Panel on Climate Change (IPCC) noted in its report "Climate Change 1995" that as computer models have improved, scientists have had to readjust radically downward their dire predictions about massive temperature increases and rises in sea levels (a supposed result of global warming). In fact, the IPCC reports that the "presence of significant errors in current models" makes most forecasts of future climate change only "fair to good."[42]

George Gilder, one of the originators of "supply side" economics and author of the 1980 classic *Wealth and Poverty*, speaks of the "irrationality of the global-warming theory." He quotes a study of the 3000-year record of thermally dependent isotope patterns in fossil sediment that shows that current global temperatures are about average, even possibly cooler than average. Known historical records demonstrate that in previous centuries temperatures were as much as 4°F higher than today—a thousand years ago temperatures were 2° warmer than today. He mentions that these clement conditions allowed the colonization of Greenland and the expansion of European populations. Actually, many anthropologists contend that Ice Age–style global freezing serves as more an impediment to social progress than global warming.

A Competitive Enterprise Institute (CEI) report offered a devastating rebuttal to the global warming thesis. This report claims that highly accurate satellite-based atmospheric temperature measurements show that the Earth's atmosphere has actually cooled by .13°C since 1979. The Arctic region of the Northern Hemisphere, which the media claim is supposedly warming, has actually gotten almost 1°C cooler in the last 50 years.

Gilder mentions that a further series of measurements shows that global temperature changes are almost entirely attributable to changes in solar, not human, activity. A 0.036 percent rise in solar intensity over the last decade unleashed an energy impact on the Earth 70 times larger than that of all the human activity put together.[43] Not surprisingly, massive cutbacks in industrial production, which global warming enthusiasts are demanding, would do little to impact global temperatures one way or the other. The CEI report reminded all parties in the debate that the Intergovernmental Panel on Climate Change itself projected that any corrective environmental policies such as the global warming treaty would have only a minuscule impact on the climate.[44]

Invariably, supporters of the global warming hypothesis will publicly trumpet this theory during both particularly temperate winters and viciously hot summers. Throughout the summer of 1998, one of the warmest on record in some Midwestern and Western sections of the United States, Al Gore, a global warming true-believer, appeared at a variety of locations to remind the local citizenry that they were experiencing the effects of global warming first hand. Texas, which was in the middle of a record heat wave, and Florida, which suffered through a summer of heat-induced forest fires, were both subjected to Gore visits, and worse, Gore speeches. The American public was unmoved—in poll after poll, in spite of the continuous summer-long drumbeat of global warming propaganda, the American people outright rejected the notion that "global warming caused a particularly bad summer." In fact, only 24 percent blamed the high heat on global warming. The majority responded that they thought it was just a bad summer.[45]

In spite of this widespread skepticism about the validity of the global warming theory, in December 1997, many countries signed the Kyoto Treaty in which they pledged to reduce their carbon-dioxide emission output to 1990s levels by 2010. The only way governments can reach such targets, many think, is to institute extremely strict guidelines requiring industry to vastly curtail production mandating people to drive less (and abandon their sports utility vehicles).

Since the signing of the treaty, many countries failed to act on the provisions. Once they realized the economic impact of the treaty, they began to rethink these draconian measures. The U.S. Senate passed a bill, the Hagel-Byrd Resolution, by a 95–0 margin, putting Clinton on notice that the Senate would not ratify any treaty that would cause serious economic harm to the United States. Fearing the treaty's defeat, the Clinton administration decided not to bring the treaty to the floor of the Senate for approval. While the treaty vote languishes, Clinton pursues a strategy of issuing regulations and Executive Orders that gradually impose some of the treaty's environmental restrictions on some areas of the United States.

Many, like the U.S. Senate, realized that the treaty's extreme measures, such as cutting back emissions to 1990 levels, would greatly harm the American and international economy. WEFA, Inc., an independent consulting group, released a study predicting that if the United States adopted the treaty's provisions, GDP per household would drop by $2000 in 2010 and cost each household a total of about $30,000 from 2001 to 2020. To pay for all the emissions costs and pass-alongs from industry, every year each household would be spending about $2500 to $5000 more than they pay for goods now. Industry would face comparable energy hikes. The Global Climate Coalition, a network of industry, union, and consumer groups opposed to the Global Warming Pact, claimed that the U.S. economy would slow considerably. Washington University's Center for the Study of American business said that the global warming pact could potentially reduce GDP by as much as 0.7 percent per year, and total employment in the United States could go down by more than 900,000. A study by DRI/McGraw-Hill stated that the United States would lose about $350 billion a year in lost output if it let emissions levels drop below 1990 levels. This would amount to a loss of about 4 percent of projected GDP. An Energy Department study of such surcharges predicts gloomy results for the auto, air transport, and semiconductor industries.[46]

Such cutbacks are unacceptable if we hope to excel in such areas as dominionization and species coalescence. In fact, we should be discovering ways to unleash the energies of the human species, not devise methods for humankind to restrict its productive activities.

However, that is the view from the expansionary side of the table. The other side has a far different agenda. Gilder relates the global warming hypothesis to a much broader zero-sum mentality. In a zero-sum game, a gain by one party necessitates a loss for another party. In economics, the system is depicted as a fixed pie that can be redistributed, but cannot be enlarged. In the end, the gains and losses always add up to zero. The zero-sum paradigm means that gains in one person's comfort and wealth must by definition lead to someone else's losses. According to Gilder, "gains for comfort and wealth of some are assumed to cause losses for other people, other species or the environment."[47] Zero-sum theory is the application of the entropy concept to the economy and the environment.

The expansionary philosophy focuses on the capabilities of systems, the "pie," as it were, to expand and progress, albeit under a very special circumstance. This condition is the presence of a consciousness that can apply its intelligence and creativity to essentially change the operation of entropy systems so that the total pie increases. Gilder obliquely refers to this process when he speaks of how human effort and imagination will expand the wealth of the system through new developments in information, food, and transportation technologies. On the broader level, we will expand the cosmic pie, enhancing the universe, endowing it with direction where there is chaos.

As we enter the third millennium, the battle between the expansionary world view and the zero-sum crowd will intensify. The expansionary camp will try to expand that pie through scientific research, technological innovation, and the enhancement of human potential; the other side will attempt to reduce output, restrain science, and restrict human activity, turning to forced redistribution of the resulting shrinking supply of goods and material.

The Clinton administration has attempted to institutionalize the zero-sum game mentality by establishing a little known but highly influential organization known as the Presidential Council on Sustainable Development. This group brings together leaders from a number of disparate sectors: businesses such as Georgia-Pacific Corp., Enron Corp., and Chevron Corp; labor unions; environmental groups such as the Natural Resources Defense Council and the Environmental Defense Fund; and representatives of national and local governments. Its goal ostensibly is to have these groups devise ways for their organizations and the nation as a whole to protect the environment without sacrificing economic growth. The language that Clinton and his supporters use to describe this Council's activities, however, betrays a hidden agenda. One of the council's major reports urged that our society take long-range steps to stabilize the country's population, including an increased outlay of federal money for family planning and contraceptive research programs. And, importantly, it says the United States, "even in the face of scientific uncertainty," should lead the world in heading off serious or irreparable global climate trends. Some of these ideas were recycled as the U.S. official position on global warming at the Kyoto conference.[48]

Many other troubling developments in the battle over growth and scientific advancement loom on the horizon. The Green Party, a staunch opponent of industrial growth, in Germany became part of a ruling coalition with Chancellor Gerhard Schroeder's Social Democrats after the 1998 elections. Almost immediately, the Greens voiced their demands for scrapping the country's entire nuclear power program and hiking the already high gasoline tax. It was only the resistance of business, labor, and the public that persuaded Chancellor Schroeder from immediately acquiescing to the Greens' demands.[49] A good many of the scientific and technological breakthroughs explored in this book have their own peculiar adversaries. For example, the Vatican, the United States, the United Nations, "concerned" scientists, and assorted ethicists proposed placing temporary bans on the cloning of humans.

Throughout 1999, a controversy brewed in Europe about whether EU countries ought to ban the import of United States–grown bioengineered food products.[50] As we entered 2000, Greens and ecologists, in conjunction with sympathizers inside the German, French, and British governments, used a variety of legal and extralegal methods to sabotage the introduction of such crops as gene-modified corn and soybeans. Some groups even suggested banning gene-modified trees pending further tests on their "safety."[51] Prince Charles, a proponent of sustainable development and a true-believer of the Gaian philosophy, stood in the forefront of the movement to ban such products. He now proclaims that he has "gone organic" when it comes to his own farming and suggests his countrymen follow suit.[52] Some in Britain criticized Charles's refusal to even consider the possibility that a potato that "has been genetically altered to make it immune to blight could be a great gift to the world." One writer suspects that his opposition to genetically modified foods reflects his unsettling view of humanity. For the "eco-prince," the author contends, "In a world fit for kings there is no room for Brazilians breeding like rats and threatening to choke nice people in eco-friendly homes by cutting down the rain forest."[53]

Let me conclude this exploration of the battle for our future by noting that such clashes are nothing new. Throughout history a torrent of controversy and suspicion greeted most new ideas and novel discoveries. So today we are witnessing a reluctance on the part of many to sanction and engage in most of the activities and projects described in this volume. Various alliances and coalitions, often backed by powerful individuals and organizations, will continue to try to prevent our species from colonizing distant spheres, modifying our genetic structure, creating clones, and taming the atom. They will exhort the human species to live in balance with nature and impact the environment and the universe as little as possible.

They will be countered by a growing intellectual movement that values growth and realizes how deleterious to the species' health are restrictions on growth and development. As the debate over the future of our species ensues, the public will discover just how close we are to unlocking the untapped potential of our species. Such a debate will generate a shared consensus over the future of the species and the means we can employ to achieve that future. To succeed in our mission, we must have full buy-in from the entire citizenry.

When the dust clears, the expansionary perspective will emerge as the real winner in this debate. I say this not because that perspective has more skillful defenders or more gifted debaters. It will triumph because we as a species have little choice but to move forward and become masters of the universe. To adopt any other view, to deny our mandate and reject our destiny, is a precondition for the extinction of our species and the stagnation of the universe.

Other species, the planet, and the universe are depending on humankind to emerge from its cultural adolescence and edge toward maturity. Let us not disappoint the cosmos!

Epilogue

At the Threshold of the Expansionary Age

The human species is about to embark on a remarkable journey to meet its mandate and achieve its destiny. Our species is challenging the constraints that nature has placed upon it and rewriting fundamental laws of the universe that only yesterday seemed immutable.

Within the next few decades, the human species will complete the task of mastering the forces of nature and will be well on its way to taking command of its evolution. We will successfully complete the processes of *dominionization, biogenesis, species coalescence,* and *cybergenesis.* We will be reprogramming our genes, commanding the behavior of the atom, and inventing the technologies that will enable us to travel at approximately the speed of light. In the process, we will become a tighter-knit, more efficient global society positioned to conquer our age-old enemies scarcity and poverty.

Our mastery of these basic processes will enable us to rapidly embark on our mission of *vitalizing* the universe. We cannot assume, however, that such progress will automatically ensure that we meet our destiny. Other events must transpire if humanity is to move forward.

For one, humanity as a whole must come to understand its destiny and accept the challenges that mandate implies. As we have seen, the emerging *expansionary* vanguard, an eclectic group of scientists, artists, futurists, and others, has already adopted this lofty vision of the human future and actively labors toward this vision's actualization. However, this vanguard represents only a small portion of the human species. We have chronicled throughout these individuals' fervent efforts to communicate this optimistic vision of the human future to the public.

Now, our educational, cultural, political, and economic institutions must also become involved in a full-throttle, turbo-charged effort to advance the human species. They can do this by facilitating economic growth in general and by helping the individual contribute to species growth. These institutions must communicate to the public an image of humankind as a species of unlimited possibilities whose potential is only beginning to be realized. For instance, schools and universities can establish general curricula and specific courses that speak to the inherent worth of the human species and its endeavors and emphasize the value of expansionary development. And all educational institutions, from grade school through college, should provide the students the intellectual tools and technological implements that will enable them to exponentially accelerate the advancement of the students' mental and creative abilities and sharpen their critical thinking skills. In this way we will unleash the potential of our students to become the inventors and innovators we sorely need to meet our species' mandate.

Government can facilitate humankind's pursuit of its destiny by becoming a champion of technological, economic, and intellectual growth. Government should strive to remove restrictions on the sciences and other institutions as they labor to expand human potential and enhance species' growth. In addition, national and international governing organizations can adopt a laissez-faire attitude toward science's efforts to help humans become a healthier and more intellectually adroit species.

In addition, the cultural establishment and its constituents must recognize its obligation to help the species advance. Artists, musicians, composers, and writers can emphasize prospecies themes in their work and help the population comprehend and appreciate all that is implied in the notion of the human future. In turn, organized foundations, as well as members of the public, must encourage artists, musicians, and cultural institutions such as museums to celebrate humanity and its natural drive toward progress and advancement.

While governments, universities, and cultural institutions can foster an environment conducive to growth and progress, it is the individual who will ultimately create the human future.

There are many ways each of us can participate in this exciting adventure. First and foremost, we can advance the species by advancing ourselves. Whenever we enrich our knowledge base or involve ourselves in activities that expand our intelligence or enhance our skills, we are advancing the species. We should endeavor to adopt an expansionary mind-set, one that is progrowth and prohuman. We must relish the opportunities for innovation and adventure and embrace the challenge to improve the environment around ourselves and the broader population.

Each of us should recognize our responsibility to maintain a sufficient level of technological literacy so that we can understand the issues affecting our lives. In addition, we must also be willing to weigh in on issues that will affect the species' ability to progress. Dozens of issues relevant to the achievement of our destiny loom on the horizon. For instance, within the next few years major debates will emerge over the introduction of a variety of new technologies and the expansion of humankind into outer space. (My next book will be entirely devoted to the emerging battles over technology and growth.)

In short, there is much that we as individuals can do to advance the species. In fact, the efforts of individuals will be the deciding factor in determining whether humanity successfully vitalizes the planet and the cosmos. Science can invent and create to its heart's content, providing us with ever-more-advanced tools for moving the species forward. However, unless we the people are willing to permit the application of such innovations to activities such as enhancing human potential and colonizing the universe, the inventions are of dubious value. We as well as our institutions ultimately must be guided by the notion that the human species' unique destiny endows it with many rights, including the rights to improve nature, live anywhere in the universe, and transform and perpetuate itself.

We stand at the threshold of the Expansionary Age, the most exciting and challenging period in the history of the human species, a golden moment in the evolution of our young universe. In this era, the human species will finally emerge onto the cosmic stage and claim its place amidst the heavens.

As we discover our destiny, we will not only change the universe, we will transform ourselves. The very act of vitalizing the universe will fortuitously lead to the fluorescence of humanity, extending us even beyond what we imagine our capabilities and powers to be. A symbiotic relationship will develop between human beings and the Humaniverse we are spawning. By rationalizing the universe, we enhance ourselves, eventually transforming ourselves into a species that will span the galaxies at the speed of light and live to ages bordering on immortality.

For centuries, philosophers and saints have been seeking the answer to a host of crucial questions: Why do we exist? What purpose do we fulfill? Why us, why here, why now? We are just now beginning to understand the nature and direction of the human species. Yet, we approach this realization with conflicting emotions bordering on both joy and trepidation. Let us hope that we overcome our apprehensions and exhibit the courage to live up to the expectations of the cosmos and the demands of our destiny.

The years, the decades, the centuries to come will be the most magnificent period imaginable—the future is an act of love. It will also be the most challenging in human history—the future is an act of will.

Imagined by our minds, created with our hands, guided by our hearts. A majestic future! The only future!

The Human Future!

Appendix

Related Web Sites

The following Web sites represent a variety of subjects and topics—technology, science, and aerospace, as well as political, economic, cultural, and social issues related to the future. These are only a starting point in your exciting exploration of a wide variety of visions of the future. But they represent a solid grounding in advanced thinking about humankind and the universe.

These Web sites can be accessed through any standard Internet provider, such as AOL, EarthLink, and Microsoft.

Zey.com

www.zey.com is the Web site of the Expansionary Institute, an organization devoted to collecting and disseminating information on human advancement and technological innovation. It is also a site where you can access photos and illustrations of the many technologies described in *The Future Factor*, including high-speed trains, mile-high cities, and medical breakthroughs such as the artificial retina. Zey.com is a way for you to keep abreast of your changing world. It looks at political, cultural, and economic factors impacting our world as we enter the brave new Expansionary Age.

At Zey.com you will find information on books, movies, and TV programs that are germane to these topics. Eventually, the Web site will become a place for an exchange of information and ideas on new technologies and more general political and economic topics.

Other Web Sites

The Astrobiology Web	http://www2.astrobiology.com/astro/
THE CATO Institute	http://www.cato.org/

Conway International	http://www.Conway.com
Davinci Institute	http://www.davinci-institute.com/
Information on Project Echelon	http://www.echelonwatch.org
Expansionary Institute	http://www.Zey.com
The Extropy Institute	http://www.extropy.com/
Foresight Institute	http://www.foresight.org/
HAARP Project	http://www.earthpulse.com/haarp/ index.html
Living Universe Foundation	http://www.luf.org/
The Mars Society	http://www.marssociety.org/
National Space Society	http://www.nss.org/
Public Broadcasting	http://www.pbs.org/science
Science Daily	http://www.sciencedaily.com/
Scientific American magazine	http://www.sciam.com/index.html
Sovereignty International	http://www.sovereignty.net/
World Federation of Great Towers	http://www.great-towers.com
World Future Society	http://www.wfs.org/

Notes

Prologue

1. Zey, Michael G. *Seizing The Future: The Dawn of the Macroindustrial Era*, Second Edition. Transaction Publishers, New Brunswick, N.J., 1998.

2. Conway, McKinley. "The Super Century Arrives." *The Futurist*, March 1998, Vol. 32, Issue 2, pp. 19–25.

3. Potter, Seth. "Microwave Power Transmission Using Tapered Beams." *Space Power*. Vol. 11, Number 2, 1992.

4. Jesdanun, Anick. "Magnetic Trains Get Federal Boost." *Associated Press*. August 1, 1998.

5. Conway, p. 22.

6. Fackler, Martin. "Japan Quietly Builds Space Programs." *Associated Press*. July 7, 1998.

7. Nash, J. Medeleine. "The Immortality Enzyme." *Time* magazine, September 1, 1997, p. 65.

8. "Artificial skin." *Popular Science*. September 1997, p. 15.

9. "The Body of the Future." Cover feature, *Popular Science*. October 1999.

10. Appleman, Philip, Editor. *Darwin: A Norton Critical Edition*. W. W. Norton and Company, New York, 1979.

11. Teilhard de Chardin, Pierre. *The Phenomenon of Man and The Vision of the Future*. Harper and Row, New York, 1975.

12. Behe, Michael J. *Darwin's Black Box: The Biochemical Challenge to Evolution*. The Free Press, New York, 1996.

13. Gribbin, John, and Martin Rees. *Cosmic Coincidences: Dark Mater, Mankind, and Anthropic Cosmology*. Bantam Books, New York, 1989.

14. Ward, Paeter D., and Donald Brownlee. *Rare Earth: Why Complex Life Is Uncommon in the Universe*. Copernicus, New York, 2000.

15. De Duve, Christian. *Vital Dust: Life as a Cosmic Imperative*. Basic Books, New York, 1995.

16. Tipler, Frank J. *The Physics of Immortality*. Anchor Books/Doubleday, New York, 1994.

17. Lerner, Eric J. *The Big Bang Never Happened*. Vintage Books, New York, 1992.

18. Kaku, Michio. *Visions: How Science Will Revolutionize the 21st Century*. Anchor Books, New York, 1997.

19. Dyson, Freeman. *Imagined Worlds*. Harvard University Press, Cambridge, Mass., 1997.

20. *Time*, April 20, 2000.

21. Hobish, Mitchell K., Ph.D., and Keith Cowing. "Astrobiology 101: Exploring the Living Universe." *Ad Astra*. January/February 1999, pp. 20–23.

Chapter 1

1. Cahill, Thomas. *The Gifts of the Jews*. Doubleday, New York, 1998, pp. 11–14.

2. Toffler, Alvin. *The Third Wave*. Bantam Books, New York, 1991.

3. Naisbitt, John. *Megatrends: Ten New Directions Transforming Our Lives*. Warner Books, New York, 1983.

4. Zey, Michael G. *Seizing The Future: The Dawn of the Macroindustrial Era*. Transaction Publishers, New Brunswick, N.J., 1998, pp. 18–32.

5. Zey, Michael G. "The Macroindustrial Era: A New Age of Abundance and Prosperity." *The Futurist*. March–April 1997, pp. 9–14.

6. Fisher, Arthur. "World's Largest Dam." *Popular Science*. August 1996, pp. 68–71.

7. "USDA to Set Up Gene Research Center." *Associated Press*. January 19, 1999.

8. Welsh, Jonathan. "Tinkering With Genes to Get a Tall, Strong 'Supertree.' " *The Wall Street Journal*. January 13, 1998, p. B1.

9. "It Takes A UN." *The Wall Street Journal*. Editorial. February 9, 1999, p. A26.

10. Ibid., p. A26.

11. "World Can Meet Food Needs Despite Population Surge." *Reuters News Service*. October 12, 1999.

12. Hertsgaard, David. *Earth Odyssey: Around the World in Search of Our Environmental Future*. Broadway Books, New York, 1999.

13. "U.S. Microbics Bugs March off to Hydrocarbon Wars." Carlsbad, Calif., *PRNewswire*. *Associated Press*. December 31, 1998.

14. Travis, John. "Meet the Superbug." *Science News.* Vol. 154, December 12, 1998, p. 376.

15. Bova, Ben. *Immortality: How Science Is Extending Your Life Span and Changing the World.* Avon Books, New York, 1998, pp. 175–182.

16. Regis, Ed. Nano: *The Emerging Science of Nanotechnology.* Little Brown, New York, 1995.

17. Drexler, K. Eric. *Engines of Creation: The Coming Era of Nanotechnology.* Anchor Books, New York, 1990.

18. Rotman, David. "Nanotech Big Money for a Small World." *MIT's Technology Review.* Vol. 101, Issue 3, May/June 1998, p. 30.

19. "News about Foresight Archive." <http://www.foresight.org/hotnews/index.html> Splash page. Update, December 5, 1998.

20. "Nanotechnology Makes Gains." *The Futurist.* June–July 1998, p. 17.

21. Rogers, Adam, and David A. Kaplan. "How We Live: The Future: Get Ready for Nanotechnology." *Newsweek.* December 2, 1997, p. 52.

22. Merkle, Ralph C. "It's a Small, Small, Small, Small World." *MIT's Technology Review.* February–March 1997, pp. 26–32.

23. Rogers and Kaplan, p. 53.

24. Stamper, Chris. "Science of the Very Small." ABCNEWS.com. January 29, 1999.

25. Rogers and Kaplan, p. 54.

26. Holland, Steve. "Clinton Offers Increase in Nanotechnology Money." *Reuters News Service.* January 21, 2000.

27. Business in Asia Today. *PRNewswire,* prepared by Asia Pulse, October 13, 1997.

28. Baldauf, Scott. "World's Oil May Soon Run Low." *The Christian Science Monitor,* September 23 1998, p. 1.

29. Ivanhoe, L. F. "Get Ready for Another Oil Shock." *The Futurist.* January-February 1997, pp. 20–23.

30. Maclean, William. "OPEC Meets to Endorse Oil Limits until April." *Reuters News Service.* September 22, 1999.

31. *The Wall Street Journal,* October 15, 1999, p. 1.

32. Mably, Richard. "IEA Warns of Big Oil Supply Deficit Next Year." *Reuters News Service.* August 10, 1999.

33. "India 99/00 Industrial Output Up 8.0 Pct." *Reuters News Service.* May 12, 2000.

34. Mitchell, Andrew. "Oil Price Surge Adds to Manufacturers' Cost Pressure." *Reuters News Service.* May 14, 2000.

35. Mankins, John C. "The Space Solar Power Option." *Ad Astra.* January/February 1998, pp. 25–26.

36. Okada, Naomi. "Japan Seeks Different Energy Sources." *Associated Press.* October 9, 1999.

37. Baker, Stephen. "For Westinghouse, A Slow Boat to China." *Business Week.* September 8, 1997.

38. "Germany to Seek Nuclear Energy End Date." *Associated Press.* October 15, 1998.

39. Energy Security: Dr. David Baldwin. *Congressional Testimony,* Research Library. America Online. October 2, 1998.

40. Wheatley, Alan. *Reuters Wire Service.* September 22, 1997.

41. Glanz, James. "Energy Research: Competition Heats Up on the Road to Fusion." *Science.* July 7, 1998.

42. Glanz, James. "Energy Research: Magnetic Fusion Researchers Think Small." *Science.* July 3, 1998.

43. "Canada Will Not Sell Nuclear Fusion Program to Iran." *Reuters News Service.* September 9, 1999.

44. Mankins, John C. "The Space Solar Power Option." *Ad Astra.* January/February 1998, pp. 25–26.

45. Lewis, John. S. *Mining the Sky: Untold Riches from the Asteroids, Comets, and Planets.* Addison Wesley Publishing Company, Reading, 1996, p. 131.

46. Pope, Gregory T. "Fly by Microwaves." *Popular Mechanics.* Vol. 172, September 1, 1995, p. 44. Also see Appell, David. "High-Power Laser Beam Launches Fuel-Less Craft." *Laser Focus World.* March 1998, Vol. 34, Issue 3, p. 90. Rivera, Rachel. "Look Ma, No Fuel!" *Science World.* October 19, 1998. UMI—ProQuest Direct Data Service.

47. Coughlin, Kevin. "Lasers Define Light Artillery." *The Newark Star-Ledger.* March 30, 1968, p. 41.

48. Muller, Judy, and Charles Gibson. "Cutting Edge." *World News Tonight with Peter Jennings.* Broadcast. August 13, 1998. Electric Library.

49. Cooper, Will, "Flying on a Beam." *Popular Science.* Vol. 253, Issue I, July 1998, p. 22. EBSCO Data Service, p. 22.

50. Farmer, Mark. "Mystery in Alaska." *Popular Science.* September 1995, pp. 79–81.

51. Rembert, Tracey C., "Discordant HAARP." (High-frequency Active Auroral Program.) Vol. 8, *E Magazine.* January 11, 1997, p. 27.

52. Ibid., p. 25.

53. Ibid., p. 26.

54. Begich, Nick, and Jeanne Manning. "Vandalism in the Sky." *www.Earthpulse.com*

55. Farmer, p. 81.

56. Biography: Nicola Tesla Biographies. <mmm.simplenet.com/frames.biographies/telsa_nikola.html>

57. Begich, Nick, and Jeanne Manning. "Angels Don't Play this HAARP: Advances in Burning Skies and Melting Minds-Project HAARP." See also *Men in Black Magazine*. Tesla Technology. Earthpulse Books. 1995. <www.meninblack.com/meninblackmag/Volume2/haarp.html>

58. Farmer, p. 80.

59. Farmer, p. 81.

60. Rembert, p. 24.

61. "EU Lacks Jurisdiction to Trace Links Between Environment and Defense." Europe Environment, Europe Information Service-Brussels, February 2, 1999. Research Library.

62. Begich and Manning. "Vandalism in the Sky."

63. Halal, William E., Michael D. Kull, and Ann Leffmann. "Emerging Technologies: What's Ahead for 2001–2030." *The Futurist*. November 1998. Electric Library.

64. Ibid., pp. 20–29.

Chapter 2

1. "Beyond 2000: 100 Questions for the New Century." *Time* cover article. November 7, 1999.

2. Begley, Sharon. "Scientists May Be Able to Predict the Fate of the Cosmos— And Draw a Genetic Blueprint That Will Forecast Your Future Health: Some Startling Breakthroughs Are Surprisingly Close." *Newsweek*. June 23, 1997.

3. Recer, Paul. "Gene Map Project Ahead of Schedule." *Associated Press*. October 23, 1998.

4. Thompson, Dick. "The Gene Machine." *Time*. January 24, 2000.

5. Thompson, Dick. "Gene Maverick." *Time*. January 11, 1999, p. 54.

6. Fisher, Lawrence M. "Alzheimer's Team Finds a New Genetic Link." *The New York Times*. April 30, 1997.

7. Leary, Warren E. "Site of Gene Tied to Parkinson's Disease Is Found." *The New York Times.* November 15, 1996.

8. Angier, Natalie. "Surprising Role Found for Breast Cancer Gene." *The New York Times.* March 5, 1996.

9. Hilts, Philip J. "A Designer Mouse Joins the Quest to Combat a Rare Genetic Disease." *The New York Times.* August 6, 1997.

10. Grady, Denise. "Brain-Tied Gene Defect May Explain Why Schizophrenics Hear Voices." *The New York Times.* January 21, 1997.

11. "Scientists Seek Hand-Tremor Gene." August 31, 1997. *Associated Press Wire Services.*

12. "Doctors Find Tumor-Fighting Gene." *Associated Press.* October 8, 1998.

13. "Glaxo Discovers Genes Behind Key Diseases." *Reuters News Service.* October 19, 1999.

14. Jaroff, Leon. "Fixing the Genes." *Time.* January 11, 1999, p. 68.

15. Grace, Eric. S. "Better Health through Gene Therapy." *The Futurist.* January-February 1998, pp. 39–42.

16. Jaroff, pp. 68–70.

17. Langretti, Robert. "Gene-Therapy Advance Is Made by Ariad and University Team." *The Wall Street Journal.* January 10, 1999, p. A14.

18. "Researchers Turn to Gene Therapy as a Possible Weapon in the Fight Against Prostate Cancer." *PRNewswire.* October 18, 1999.

19. Goldberg, Jeff. "Gene Therapy." *Life Magazine.* Fall 1998, p. 70.

20. Begley, Sharon. "Lifestyle" Section. *Newsweek.* November 9, 1998. Electric Library.

21. Fackelmann, Kathleen. "It's a Girl! Is Sex Selection the First Step to Designer Children?" *Science News.* November 28, 1998. Electric Library.

22. Miller, Henry I. "Better Genes for Better Living." *The Wall Street Journal.* August 21, 1999, p. B1.

23. Lemonick, Michael D. "Designer Babies: Parents Can Now Pick a Kid's Sex and Screen for Genetic Illness: Will They Someday Select for Brains and Beauty Too?" *Time.* January 11, 1999, pp. 64–66.

24. Nash, J. Medeleine. "The Immortality Enzyme." *Time.* September 1, 1997, p. 65.

25. "Cloning of Human Telomerase Gene Reported in Science; Telomerase Plays a Key Role." *Associated Press.* August 14, 1997.

26. Bova, p. 133.

27. "Geron Stock Soars after Enzyme Research." *Reuters.* December 29, 1998.

28. Merkle, Ralph C. "Nanotechnology and Medicine." Originally published in *Advances in Anti-Aging Medicine*, Vol. I, edited by Dr. Ronal M. Klatz, Lievert press, 1996, pp. 277–286. <http://nano.xerox.com/nanotechAndMedicine.html>

29. Bova, Ben. *Immortality: How Science Is Extending Your Life Span and Changing the World.* Avon Books, New York, 1998, pp. 176–182.

30. Lampton, Christopher, *Nanotechnology Playhouse: Building Machines from Atoms.* Waite Group Press, California, 1993, p. 75.

31. Regis, Ed. *Nano: The Emerging Science of Nanotechnology.* Little Brown, Boston, 1993, p. 6.

32. Merkle, p. 278.

33. Uehling, Mark D. "Counting 6LL3." *Popular Science.* May 1997, pp. 74–75.

34. "U.N.: Human Cloning Unacceptable." *Associated Press.* April 23, 1997.

35. Bashi, Sari. "Israel Bans Genetic Cloning." *Associated Press.* December 30, 1998.

36. Reilley, Matthew. "Man Who'd Clone Others Generates Scorn." *Newark Star-Ledger.* January 8, 1998, p. 14.

37. "S. Korean Team Reports Cultivating Human Embryo, Reuters Says." *PRNewswire*, December 17, 1998.

38. Fox, Maggie. "Korean Claim of Human Cloning Raises Urgent Issues." *Reuters News Service.* December 17, 1998.

39. "American Life League President Judie Brown Issued the Following." *PRNewswire.* December 17, 1998.

40. "Cloning Explained." *Associated Press.* April 12, 1997.

41. Hall, Alan. "The Genetic Key to Turning Back the Biological Clock." *Business Week Online.* May 10, 2000.

42. "Cloning Breakthrough Detailed on Life Extension Web Site." *PRNewswire.* April 28, 2000.

43. Giri, Priya. "Cloning Organs." *Life Magazine.* Fall 1998, p. 56.

44. "China Bans Human Cloning." *Associated Press.* May 12, 1997.

45. Kurtenbach, Elaine. "Chinese Mull Test-Tube Pandas." *Associated Press.* July 22, 1997.

46. "Thai Scientists to Clone Elephant." *Associated Press.* January 2, 1999.

47. Feinsilber, Mike. "Cloning." *Associated Press.* April 4, 1997.

48. Crews, Christian. "Future View." *The Futurist*, June-July 1998, p. 72.

49. Goldberg, Jeff. "Artificial Heart." *Life Magazine*, Fall 1998, p. 85.

50. "Artificial Skin." *Popular Science.* September 1997, p. 15.

51. McVicar, Nancy. "Infant Receives Life-Saving Bioengineered Skin." *Newark Star-Ledger*. January 12, 1999, p. 4.

52. Hirshberg, Charles. "The Body Shop." *Life Magazine*. Fall 1998, pp. 51–56.

53. Lemonick, Michael. "Tomorrow's Tissue Factory." *Time*. January 11, 1999, p. 89.

54. "Human Stem Cell Breakthrough Heralds New Era." *Associated Press News Service*. November 5, 1998.

55. "Experiment Shows Regrowth of Spinal Cord Fibers." *Reuters News Service*. April 28, 2000.

56. Silver, Brian. *The Ascent of Science*. Oxford University Press, Oxford, England, 1997, p. 273.

57. Wingerson, Lois. *Unnatural Selection: The Promise and the Power of Human Gene Research*. Bantam Books, New York, 1998, p. 329.

Chapter 3

1. "Techno Sapiens." Cover inscription. *Time Digital*. April 27, 1998.

2. Buechner, Maryanne Murray. "Techno Sapiens." *Time Digital*. April 27, 1998, p. 30.

3. Rose, Michael, "Artificial Retina." *Life Magazine*. Fall 1998, p. 67.

4. Coughlin, Kevin. "Eyes of the Future." *The Newark Star-Ledger*. June 3, 1996, p. 21.

5. Smith, Jack, and Peter Jennings. "Cutting Edge." *ABC World News Tonight with Peter Jennings*. November 11, 1998.

6. Kurzweil, Ray. *The Age of Spiritual Machines*. Viking, New York, 1999, pp. 65–70 and pp. 127–128.

7. O'Malley, Chris. "The Binary Man: Step One." *Popular Science*. March 1999, p. 64.

8. Chartier, John. "Mind Meets Machine." *Morristown Daily Record*. March 17, 1998, p. A1.

9. Lange, Larry. "Chip Implants: Weird Science with a Noble Purpose." *Electronic Engineering Times*. February 10, 1997, Issue 940, p. 24.

10. Ibid., p. 25.

11. Ibid., p. 25.

12. Kurzweil, Ray, pp. 65–70.

13. Moravec, Hans. *Robot: Mere Machine to Transcendent Mind.* Oxford University Press, Oxford, England, 1999, p. 60.

14. Johnson, George. "Giant Computer Virtually Conquers Space and Time." *The New York Times.* 1997.

15. "IBM Supercomputer to Assess Mankind's Impact on Earth's Climate." *Associated Press Business Wire.* August 11, 1999.

16. O'Malley, Chris. "Biology Computes." *Popular Science.* February 1999, pp. 61–65.

17. Taubes, Gary. "Evolving a Conscious Machine." *Discover,* June 6, 1998, pp. 72–80.

18. Kurzweil, p. 221.

19. Kurzweil, pp. 277–280.

20. Moravec, Hans. *Robot: Mere Machine to Transcendent Mind.* Oxford University Press, Oxford, England, 1999.

21. Moravec, pp. 93–96.

22. Norman, Donald A. "The Melding of Mind and Machine." *MIT Technology Review.* April 1997, pp. 29–32.

23. Kurzweil, p. 160.

Chapter 4

1. Davidson, Frank P., with John Stuart Cox. *Macro: A Clear Vision of How Science and Technology Will Shape Our Future.* William Morrow and Company, New York, 1983, p. 197.

2. Conway, McKinley. "The Super Century Arrives." *The Futurist.* March 1998, p. 20.

3. McKinley, p. 22.

4. Pohl, Frederick. "Disappearing Technologies: The Uses of Futuribles." *The Futurist.* February 1999, pp. 30–35.

5. "Ahead by a Nose." *Popular Science.* July 1997, p. 14.

6. "Track Record: 279.6 mph in Japan." *Associated Press Wire Service.* October 4, 1997.

7. "250-Mph Subway." *Popular Science.* August 1987, p. 27.

8. "Amtrak Prepares for High-Speed Rail." *United Press International.* January 28, 1999.

9. Johnson, Glen. *"High-Speed Rail Service Faces Roadblocks." Associated Press.* January 26, 1999.

10. Jesdanun, Anick. "Magnetic Trains Get Federal Boost." *Associated Press.* August 10, 1998.

11. Jesdanun, Anick. "Fast Train Fund Finalists Named." *Associated Press.* May 22, 1999.

12. Nakarmi, Laxmi, and William J. Holstein. "High Stakes Victory of Runaway Train." *Business Week,* December 5, 1994. Laxmi Nakarmi and William J. Holstein in Seoul, with Farah Nayeri in Paris and Karen Lowry Miller in Munich.

13. Quinn, Andrew. "Once a Sci-Fi Dream, Skycar Nears Reality." *Reuters News Service.* October 10, 1999.

14. Ibid.

15. Sweetman, Bill. "The 21st Century SST." *Popular Science.* April 1998, pp. 56–60.

16. "X-33 Metallic Heat Shield 'Ready For Flight.' " *Regulatory Intelligence Data.* February 3, 1999. Electric Library.

17. Larson, Ruth. "NASA Seeks Shuttle Alternative; Venturestar's Goal to Reduce Launch Costs." *The Washington Times.* November 29, 1998, p. D8. Electric Library.

18. Longman, Philip, et al. "The World Turns Gray." *U.S. News and World Report.* March 1, 1999.

19. Garten, Jeffrey. *The Big Ten: The Big Emerging Markets and How They Will Change Our Lives.* Basic Books, New York, 1997, pp. 3–23.

20. *Business in Asia Today. PRNewswire* prepared by Asia Pulse. August 14, 1997.

21. "Indian Gov't. Forecasts 5.8 Pct. Growth in Economy." "Malaysian Auto Sales Likely to Rise 12 Pct. in 1999." Summary of *Business in Asia Today. PRNewswire.* February 11, 1999.

22. Garten, pp. 33–41.

23. "Royal Group of Canada to Set Up Factories in Philippines." Summary of *Business in Asia Today.* February 2, 1999. *PRNewswire* prepared by *Asia Pulse.*

24. *Business in Asia Today. PRNewswire* prepared by *Asia Pulse.* August 13, 1997.

25. "Report: China to Invest in Mining." *Associated Press Wire Service.* September 7, 1997.

26. "Chinese Bank to Grant $3.6 Bln Loans for Powr Generation." *Business in Asia Today. PRNewswire.* February 2, 1999.

27. "The Road from Imitation to Innovation." *The Economist.* May 18, 1996, pp. 80–81.

28. Huang, Annie. "Taiwan Becomes Computer Giant." *Associated Press Wire Service.* September 7, 1997.

29. "India: A High-Tech Success." *The Wall Street Journal.* August 15, 1997, pp. B6–B10.

30. Cardwell, Donald. *The Norton History of Technology.* W. W. Norton and Company, New York, 1995.

31. Dalton, Richard J., Jr. "Netting New Crop of Web Surfers: Novices Are Altering What's Hot and What's Not in Cyberspace." *Newsday.* January 15, 1999, p. A64.

32. "China Says Internet Users Surge to 1.5 Million." *Reuters News Service.* January 15, 1999.

33. "Internet's Latest 'Great Satan.' " *Reuters News Service.* January 22, 1999.

34. Ibid.

35. Dalton, 1999. Some feel that the Internet and other such global communications technologies possess a decided dark underbelly. Intelligence agencies such as the NSA and Britain's Defense Signals Directorate DSD have very quietly established "Project Echelon," a network of powerful computers capable of voice recognition that can monitor every international phone call, fax, e-mail, and radio transmission. These supercomputers ostensibly home in on a long list of keywords in these messages looking for evidence of unlawful behavior on the part of the messages' senders or receivers. Many consider the program's methodology, arbitrarily poring over all messages and e-mail it intercepts, tantamount to illegally spying on private citizens. The security agencies of all governments involved in the program deny its very existence. The NSA, which by its charter is answerable only to the U.S. President, has refused to turn over any Echelon-related documents to Congressional committees attempting to investigate the matter. By 2000, the American Civil Liberties Union was introducing court action to pressure various U.S. agencies to admit the existence of Echelon and reveal the nature and extent of the program. (See www.echelonwatch.org.)

36. Pohl, Frederick. "Disappearing Technologies: The Uses of Futuribles." *The Futurist.* February 1999, pp. 30–35.

37. Zey, pp. 278–283.

38. "The Virtual Nose at Montefiore Medical Center Makes Learning High-Risk Surgery Safer." *PRNewswire.* January 21, 2000.

39. Lo, Catharine. "Space Jam." Archive feature. *Wired Digital.* October 1998. <www.wired.com>

40. "China Launches Satellites with Brazil." *Reuters News Service.* October 14, 1999.

41. "Commercial Satellite Reaches Orbit." *Associated Press.* October 10, 1999.

42. "Ariane Rocket Puts Loral Satellite into Orbit." *Reuters News Service.* September 25, 1999.

43. Taggart, Stewart. "Rocket Change: Got a Satellite? Have We Got a Launch Vehicle for You." Archive Feature. *Wired Digital.* October 1998. <www.wired.com>

44. Heylighen, F. "The Superorganism and Its Global Brain." Comment by Don Edward Beck. *Principia Cybernetica Web.* July 7, 1997. <pespmc1.vub.ac.be/SUPORGLI.html>

45. Ibid.

46. Hubbard, Barbara. "Conscious Evolution: Examining Humanity's Next Step." *The Futurist.* September–October 1993, pp. 38–41.

47. Ibid., p. 40.

48. Teilhard de Chardin, Pierre. *The Phenomenone of Man.* Harper and Row, New York, 1975, pp. 287–288.

49. "Soul Catcher Implants." *Electronic Telegraph* (Online edition of *The Daily Telegraph, the Daily Mail*). July 18, 1996. Electric Library.

50. Rixmer, Rob. "The Melding of Mind and Machine." *The New York Times.* August 10, 1998, pp. B1.

51. "The Future of Cybernetics." *ABC World News Tonight.* August 6, 1999. <www.abcnews.com/onair/DailyNews/Chat_Cochrane990806.html>

52. Teilhard de Chardin, p. 138.

53. Ibid., pp. 239–240.

Chapter 5

1. "China Launches First Space Shuttle—Xinhua." *Associated Press.* November 22, 1999.

2. Savadove, Bill. "Interview: China to Put Man in Space 'Soon.'" *Reuters.* November 29, 1999.

3. Schefter, Jim. "Reaching for the Rings." *Popular Science.* October 1997, pp. 61–65.

4. Dunn, Marcia. "NASA to Send Signatures to Saturn." *Associated Press.* September 22, 1997.

5. Dunn, Marcia. "NASA about To Launch Deep Space 1." *Associated Press.* October 18, 1998.

6. "Satellite Launched to Study Stars." *Associated Press.* March 4, 1999.

7. Zey, Michael G. "The Macroindustrial Era: A New Age of Abundance and Prosperity." *The Futurist*, March/April 1997, pp. 9–14.

8. Fackler, Martin. "Japanese Mission to Mars Set Date." *Associated Press*. June 30, 1998.

9. "Mars Remains Hopeless Until 2003: Japans 'Nozomi' (hope) Probe Delayed." *Ad Astra*. March/April 1999, p. 8.

10. "Moon Missions May Make Comeback." *Associated Press*. August 28, 1997.

11. Pierce, Anne. "A New Initiative for a New Millennium." *Ad Astra*. March/April 1999, p. 5.

12. Ibid.

13. Steen, Michael. "Putin Supports Mir, International Space Lab." *Reuters News Service*. April 12, 2000.

14. "Mega Science Projects Follow Millennium." *United Press International*, January 23, 1999. Electric Library.

15. F2000 Va-HUD-Appropriations: Daniel S. Goldin, Congressional Testimony, March 18, 1999. *Federal Document Clearinghouse*. Electric Library.

16. "Mir's Crew Restarts Computer." *Associated Press*. September 9, 1997. "Soyuz Prepares for Mir Mission." *Associated Press*. August 3, 1997.

17. "Russia's Mir Unmanned, Possibly for Good." *Associated Press*. August 27, 1999.

18. "Historic Docking of Privately Backed Mission to Mir." April 6, 2000.

19. *Frommer's The Moon: A Guide for First-Time Visitors* (Frommer Other) 1999.

20. Elboghdady, Dina. "Firms Seek Tourism's New Frontier." *KRT News Service*. September 15, 1997.

21. Lauer, Charles J. "Places in Space," pp. 24–28. Diamonds, Peter H. "Space Tourism . . . I Believe. Do You?" *Ad Astra*. March-April 1996, pp. 34–37.

22. Futterman, Matthew. "Race Is On to Produce Giant Leap for Tourists." *Newark Star-Ledger*. June 12, 1998, p. C 1.

23. Ibid., p. C1.

24. Hobish, Mitchell K., and Keith Cowing. "Astrobiology 101: Exploring the Living Universe." *Ad Astra*. January/February 1999, pp. 20–23.

25. Goldin, Congressional Testimony.

26. Zubrin, Robert, *The Case for Mars: The Plan to Settle the Red Planet and Why We Must*. The Free Press, New York, 1996, pp. 223–225.

27. Raeburn, Paul. "Manned Mission to Mars." *Popular Science*. February 1999, p. 44.

28. Chaikin, Andrew. "Back at Mars." *Popular Science*. October 1997, p. 64.

29. Schmidt, Greg, and Mike Hawes. "Robots vs. Humans in Space: Both Will Be Required." *Ad Astra*. January/February 1999, pp. 41–43.

30. Leifer, Stephanie. D., "Reaching for the Stars"; Harris Henry M., "Light Sails"; McClinton, Charles R. "Air-Breathing Engines"; Beardsley, Tim. "The Way to Go in Space"; Powell, James R. "Compact Nuclear Rockets." *Scientific American*, February 1999. Electric Library.

31. Raeburn, pp. 40–48.

32. Ober, James. "Missionaries to Mars." *Technology Review*. January 1, 1999, pp. 54–59.

33. Zubrin, Robert, *The Case for Mars*.

34. McKay, Christopher P. "Bringing Life to Mars." *Scientific American*, March 1999.

35. Tierney, John "Martin Chronicle." *Reason*. February 1999, pp. 24–33; "Making Planets People-Friendly." *The Futurist*, March/April 1996; Sheffield, Charles, "The Greening of Mars." *The World & I*. February 2, 1998, p. 180. Electric Library.

36. Foster, David. "Flight Took Wing in 20th Century." *Associated Press News Service*. November 20, 1999.

Chapter 6

1. Lytkin, Vladimir V. "Tsiolkovsky's Inspiration." *Ad Astra*. November/December 1998, pp. 35–39.

2. Ibid., p. 39.

3. Esfandiary, F. M. *Up-Wingers: A Futurist Manifesto*. Popular Library, New York, 1977. Back-cover endorsement.

4. Esfandiary, F. M. *Up-Wingers: A Futurist Manifesto*.

5. Dyson, Freeman. *Imagined Worlds*. Harvard University Press, Cambridge, Mass., 1997, p. 167.

6. Kaku, Michio. Visions: *How Science Will Revolutionize the 21st Century*. Anchor Books, New York, 1997.

7. Savage, Marshall. *The Millennial Project: Colonizing the Galaxy in Eight Easy Steps*. Empyrean Publishing Limited, Denver, 1992.

8. Wright, Robert. Nonzero: *The Logic of Human Destiny*. Pantheon Books, New York, December 1999.

9. Morris, Simon Conway. "Where Are We Headed? Robert Wright Argues That

Human History Does Indeed Have a Purpose." *The New York Times Book Review.* January 30, 2000, p. 1.

10. Ferris, Timothy. *The Whole Shebang: A State of the Universe Report.* Simon and Schuster, New York, 1997, pp. 290–300.

11. Cowen, Ron. "The Greatest Story Ever Told." *Science News.* December 19 & 26, 1998, pp. 392–395.

12. Glynn, Patrick. God: *The Evidence.* Prima Publishing, Rocklin, Calif., 1997, pp. 21–55.

13. Kaku, Michio. "What Happened Before the Big Bang?" *Astronomy.* May 1996, pp. 34–42. Electric Library.

14. Ibid.

15. Ferris, p. 301.

16. Gould, Stephen Jay. *Dinosaur in a Haystack.* Harmony Books, New York, 1996. Quoted in Jonathan Wells, "The Retreating Revolutionary." *The World & I.* March 1, 1996, p. 270.

17. Wells, Jonathan, "Issues in the Creation-Evolution Controversies." *The World & I.* January 1, 1996, p. 294. Electric Library.

18. Schroeder, Gerald. *The Science of God.* The Free Press, New York, 1997, p. 116.

19. Ibid., Chapters 6 and 7.

20. Ibid., p. 112.

21. Lipkin, Richard. "Unicellular Organisms." *Science News.* July 23, 1994, pp. 58–59.

22. Ibid., p. 59.

23. Kauffman, Stuart. *At Home in the Universe.* Oxford University Press, New York/Oxford, England, 1995, p. 8.

24. Bailey, Ronald, "Origin of the Specious: Why Do Neoconservatives Doubt Darwin?" *Reason.* July 1, 1997, pp. 22–27.

25. Easterbrook, Gregg. "Science Sees the Light." *New Republic.* October 12, 1998, pp. 24–30.

26. Ornstein, Robert. *The Evolution of Consciousness: Of Darwin, Freud, and Cranial Fire: The Origins of the Way We Think.* Touchstone, New York, 1991, p. 40.

27. Ibid.

28. Dyson, *Imagined Worlds*, p. 172.

29. Dyson, Freeman. "Time Without End: Physics and Biology in an Open Universe." *Reviews of Modern Physics.* 1979, Vol. 5, pp. 447–460.

30. Gribbin, John, and Martin Rees. *Cosmic Coincidences: Dark Mater, Mankind, and Anthropic Cosmology.* Bantam Books, New York, 1989, p. 291.

31. Quoted in Easterbrook, p. 27.

32. Meyer, Michael. "Ex Astra: Life from the Stars." *Ad Astra.* January/February 1999, pp. 28–31.

33. de Duve, Christian. *Vital Dust: Life as a Cosmic Imperative.* Basic Books, New York, 1995.

Chapter 7

1. Zey, Michael G. *Seizing the Future: The Dawn of the Macroindustrial Era.* Transaction Publishers, New Brunswick, N.J., 1998, pp. 433–436.

2. Gribbin, John, and Simon Goodwin. *Origins: Our Place in Hubble's Universe.* The Overlook Press, Woodstock, N.Y., 1998, p. 116.

3. DiChristina, Mariette. "Star Travelers." *Popular Science.* June 1999, pp. 54–59.

4. Hickam, Homer H. "Bring Back the Rocket Boys." *The Wall Street Journal.* May 21, 1999, p. A12.

5. DiChristina, p. 55.

6. Ibid., p. 56.

7. Savage, Marshall. *The Millennial Project: Colonizing the Galaxy in Eight Easy Steps.* Empyrean Publishing Limited, Denver, 1992.

8. Savage, pp. 416–422.

9. "Gary Klein Studies How Our Minds Dictate Those 'Gut Feelings.' " *The Wall Street Journal.* August 7, 1998, p. B1.

10. Chiao, Raymond Y., Paul G. Kwiat, and Aephraim M. Steinberg. "Faster Than Light?" *Scientific American.* August 1993, pp. 52–60.

11. Watson, Andrew, "Physicists Create Trio of Entangled Particles." *ScienceNOW,* February 22, 1999. Electric Library.

12. Neimark, Jill, Tracy Cochran, and Larry Dossey. "Nature's Clones." *Psychology Today* cover story. Vol. 30, July 1, 1997, p. 36(15). Electric Library.

13. Silver, Brian L. *The Ascent of Science.* Oxford University Press, Oxford, England, 1997, pp. 446–447.

14. Savage, p. 247.

15. Dyson, Freeman. *Imagined Worlds.* Harvard University Press, Cambridge, Mass., 1997, p. 171.

16. Veon, Joan. *Prince Charles: The Sustainable Prince.* Hearthstone Publishing Ltd., Oklahoma City, 1998.

17. Savage, p. 23.

Chapter 8

1. Irwin, Jim. "Environmental Group Released Minks." *Associated Press.* October 29, 1998.

2. Dyer, Gwynne. "The Community of Equals." *The Washington Times.* April 5, 1999, p. A15.

3. Lofton, Bill and Carol. "New Zealanders Fight for Apes' Rights." *Science.* February 1998.

4. Uhlig, Robert. "Ape of Things to Come." *The Daily Telegraph.* February 11, 1999. Electric Library/America Online.

5. Trefil, James. *Are We Unique: A Scientist Explores the Unparalleled Intelligence of the Human Mind.* John Wiley and Sons, New York, 1997, pp. 56–60.

6. Kaku, Michio. *Visions: How Science Will Revolutionize the 21st Century.* Anchor Books, New York, 1997, p. 98.

7. Lofton, Bill and Carol, op. cit.

8. Dyer, op. cit.

9. Ibid.

10. Kurzweil, Ray. *The Age of Spiritual Machines Viking,* New York, 1999, p. 90.

11. Penrose, Roger. *The Emperor's New Mind.* Oxford University Press, New York, 1989; *Shadows of the Mind.* Oxford University Press, New York, 1994.

12. Freedman, David. *The Brainmakers.* Simon and Schuster, New York, 1994, pp. 184–188.

13. At a 1999 meeting of the World Future Society I had the pleasure of meeting a British Telecom resident futurist, Dr. Ian Pearson. We had a brief discussion about the topic of "machine intelligence," and I mentioned to Dr. Pearson Penrose's contention that machines will never achieve true intelligence or consciousness. Dr. Pearson reacted strongly to Penrose's books, according no credence whatsoever to Penrose's theories. (I believe that the word *rubbish* was applied to Penrose's two books.) In fact, Pearson told me that he was certain smart machines would approximate human consciousness, within at most a few decades. (Interview of Dr. Ian Pearson, July 29, 1999. World Future Society conference, Washington, D.C.)

14. Norman, Donald A. "Melding Mind and Machine." *Technology Review*. April 1997, pp. 29–32.

15. Kurzweil, op. cit., p. 280.

16. Moravec, Hans. *Robot: Meere Machine to Transcendent Mind.* Oxford University Press, Oxford, England, 1999. Chapter 5, "The Age of Robots."

17. Cochran, Peter. "Death by Installments." *Electronic Telegraph*. December 31, 1996. Electric Library.

18. Kurzweil, op. cit.

19. Cochrane, op. cit.

20. Kurzweil, Ray. "Spiritual Machines: The Merging of Manna and Machines." *The Futurist*. November 1999, pp. 16–22.

21. Hargrove, Thomas, and Joseph Bernt. "The Next 1000 Years: Americans Predict the Future." *Scrippps Howard News Service*. November 30, 1999.

22. Harmon Amy, "Alien Craze: As 50th Anniversary of Roswell Nears, More Americans Are UFO Faithful." *Minneapolis Star Tribune*. July 20, 1997, p. 19A.

23. Ethnic NewsWatch. "Outlandish." *Brazzil.* July 31, 1997. Electric Library.

24. Recer, Paul. "Scientists Eye Life Beyond Earth." *Associated Press*. October 14, 1998.

25. Sagan, Carl. *Pale Blue Dot.* Random House, New York, 1994, pp. 351–363.

26. Nieves, Evelyn. "Space Detective: First Man to Hold Academic Chair on Search for Extraterrestrial Life Says Post Is Welcome Recognition That His Work Is Serious Endeavor." *The Dallas Morning News, The New York Times News Service*. April 23, 1999, p. 47A. Electric Library.

27. Seabrook, Charles. "Aliens' Visits Unlikely, Expert Says Extraterrestrials Would Be Bound by the Same Laws of Physics We Are, Says Physics Professor." *The Atlanta Constitution*. March 25, 1999, p. E11. Electric Library.

28. Nieves, op. cit., p. 47.

29. Dave, Abe. "Flying Saucers: The Real Story." *Popular Mechanics*. January 1995, pp. 50–55. Electric Library.

30. Hadnot, Ira J., "Norman Remley." *The Dallas Morning News*. September 27, 1998, p. 1J.

31. Wynar, Roahn. "Tenured Fruitcakes Should Go." *Daily Texan* via U-Wire. University Wire, July 10, 1998.

32. Hargrove, Thomas, and Joseph Bernt, op. cit.

33. Easterbrook, op. cit., pp. 24–30.

34. Turner, Frederick. "Great Expectations." *Chronicles*. May 1995, pp. 30–32.

35. Sagan, op. cit., p. 371.

36. Philips, Lisa E. "Love, American Style Blame It on the Boomers: Generation X Is Making Marriage Work for Them." *American Demographics*. February 1999, pp. 56–57.

37. Wolfson, Adam. "Life Is a Gamble." *The Wall Street Journal*. August 14, 1998, p. W11.

38. Ibid.

39. Lerner, Eric. *The Big Bang Never Happened*. Vintage Books, New York, 1991.

40. Lerner, Eric. "Natural Science at the Edge Assessing the Pillars of Cosmology." *World & I*. January 1994.

41. Peterson, Ivars. "State of the Universe: If Not with a Big Bang, Then What?" *Science News*. April 13, 1991, pp. 232–236.

42. Merline, John. "Global Warming's Cooling Off as Science Improves, Gloomy Forecasts Fade." *Investor's Business Daily* (National Issue). October 6, 1997.

43. Gilder, George. "Zero-Sum Folly, from Kyoto to Kosovo." *The Wall Street Journal*. May 6, 1999, p. 31.

44. Balling, Jr., Robert C. "Global Warming: Messy Models, Decent Data, and Pointless Policy" in *The True State of the Planet*, Ronald C. Bailey, ed. The Free Press, New York, 1995.

45. *The Wall Street Journal*. July 31, 1998, p. A1.

46. "The Greenhouse Treaty's Chilling Prospect." (Editorial) *Investor's Business Daily*. October 6, 1997.

47. Gilder, op. cit., p. 31.

48. "AEP Chief Draper Appointed to President's Council on Sustainable Development." *PRNewswire*. February 22, 1999.

49. Geitner, Paul. "Germany to Seek Nuclear Energy End." *Associated Press*. October 15, 1998.

50. "Poll: Brits Want Modified Food Ban." *Associated Press*. February 21, 1999.

51. Mann, Michael. "EU's Top Judges Asked to Rule on Landmark GMO." *Reuters News Service*. November 9, 1999. "Green Group Warns on GM Tree Development." *Reuters News Service*. November 9, 1999. "EU May Vote on Friday Whether to Approve New GMOs." *Reuters News Service*. October 28, 1999.

52. "No Need to Fear Mutant Peaches." *The Wall Street Journal*. June 22, 1998. <www.Monsanto.co.uk>

53. de Lisle, Leanda. "Listen Up, Prince Charles." *The Spectator*. London. August 28, 1999. I have encountered stories about the various reasons Prince Charles supports ecological and environmental movements. Partly, his adherence to the

Gaian philosophy strongly influences his antitechnological views. However, in a discussion with Joan Veon, a respected researcher of the influence of the Royal Family on world affairs, I learned that the Royals' concept of the "special place" they hold in the human hierarchy conditions their views on everything else, including the environment. According to Veon, the British royal family actually sees the world in feudal terms: the Earth is "their planet," and we are "their tenants." Charles's adherence to such concepts as sustainable development, zero-population growth, and Gaia makes perfect sense, then. He wants to make sure that the rest of us behave and act as responsible stewards while we live on his (royal) family's manor. (Based on a conversation author had with Joan Veon, July 29, 1999, Washington, D.C.)

Index

About the Author

Dr. Michael G. Zey, considered one of the boldest and most exciting futurists on the contemporary scene, consults to high-technology and Fortune 500 corporations as well as to various government agencies on future planning. He speaks regularly and extensively about the future on the international lecture circuit, and has appeared on *The Wall Street Journal Report*, CNN's *Business Report*, CNBC's *America's Vital Signs*, and *The Turning Point*. His ideas and articles on societal trends have appeared in such diverse publications as *Computer Decisions, Le Monde, The Wall Street Journal, Forbes, Training and Development*, and *Investor's Business Daily*, as well as the Gannett and Scripps-Howard wire services. Zey holds a Ph.D. in sociology, and is executive director of the Expansionary Institute (www.zey.com). The author of numerous books, including *Seizing the Future: The Dawn of the Macroindustrial Era* and *The Mentor Connection*, Zey is a professor at Montclair State University, Upper Montclair, NJ.